Exercises in Rethinking Innateness

Neural Network Modeling and Connectionism
Jeffrey L. Elman, Editor

Exercises in Rethinking Innateness

A Handbook for Connectionist Simulations

Kim Plunkett &
Jeffrey L. Elman

A Bradford Book
The MIT Press
Cambridge, Massachusetts
London, England

Printed and bound in the United States of America.

ISBN 0-262-66105-5

Second printing, 1998

"This is dedicated to the ones we love."

Contents

viii   _Contents_

Series foreword

The goal of this series, Neural Network Modeling and Connectionism, is to identify and bring to the public the best work in the exciting field of neural network and connectionist modeling. The series includes monographs based on dissertations, extended reports of work by leaders in the field, edited volumes and collections on topics of special interest, major reference works, and undergraduate and graduate-level texts. The field is highly interdisciplinary, and works published in the series will touch on a wide variety of topics ranging from low-level vision to the philosophical foundations of theories of representation.

Jeffrey L. Elman, Editor

Associate Editors:

James Anderson, Brown University
Andrew Barto, University of Massachusetts, Amherst
Gary Dell, University of Illinois
Jerome Feldman, University of California, Berkeley
Stephen Grossberg, Boston University
Stephen Hanson, Princeton University
Geoffrey Hinton, University of Toronto
Michael Jordan, MIT
James McClelland, Carnegie-Mellon University
Domenico Parisi, Instituto di Psicologia del CNR
David Rumelhart, Stanford University
Terrence Sejnowski, The Salk Institute
Paul Smolensky, Johns Hopkins University
Stephen P. Stich, Rutgers University
David Touretzky, Carnegie-Mellon University
David Zipser, University of California, San Diego

Preface

Learning to do simulations is like learning many skills. You learn best by doing. Reading about simulations and watching others do them can be useful. But when it comes down to it, there is no substitute for first-hand experience.

Connectionists are certainly not the first to develop models, nor the first to run simulations of them. But simulations play a particularly important role in connectionism. Why is this?

Connectionist models are often deceptively simple. After all, they are composed of simple neuron-like processing elements (usually called "nodes"). In fact, the nodes of most connectionist models are vastly simpler than real neurons. There appear to be no elaborate rules nor grammars (which has led some to argue—erroneously in our opinion—that connectionist models "have no rules"). The nodes are simply hooked together in rather straightforward arrangements, presented with some stimulus, and then allowed to interact freely. Periodically, some nodes may output values.

Put this way, one might wonder why go to the bother of writing a computer program to actually simulate a process whose outcome ought to be entirely predictable in advance. The surprise is that very often the outcome is *not* what one might have predicted. The dynamics of these simple systems turn out to be quite complex. For various reasons, including the fact that they are nonlinear (see Chapter 4 in the companion volume, *Rethinking Innateness*), it is frequently only possible to know how they will perform by running careful experiments and observing their behavior under controlled conditions. (In this respect, they are not unlike humans and other unpredictable biological systems.) Simulations are therefore crucial, and modelers' expectations are often betrayed.

The experimental aspect of running simulations is something we want to emphasize in this book. For many of our readers, particularly those who have been trained as developmentalists, the experimental methodology will be familiar. Rather than thinking of doing simulations as a very new and different way of doing research, we encourage our experimentalist readers to regard this as simply doing experiments with a different sort of subject pool!

There is also an important conceptual role which is played by simulations, and we wish to emphasize this as well. Although it may be satisfying to build a model which successfully learns some behavior, especially if that behavior is complex and not obviously learnable, we feel the real value of simulating models is that we can then ask, "How did the model solve the problem?" When we run experiments with human subjects, the most we can usually do is to construct theories which are consistent with our data. This is an inferential process. At best, we may have multiple sources of evidence which converge on the same theory. But rarely do we have the opportunity to look inside our subjects' heads and verify that the mechanism responsible for the behavior we have observed is indeed at work.

Simulated models, on the other hand, *can* be opened up for inspection and analysis. What we often find is that the mechanism which is actually responsible for the behavior we observed from the outside looks very different from what we guessed it might be. We must be careful, of course, not to assume that the model's solution is necessarily the same as the one discovered by species or children. The model can be a rich source of new hypotheses, however. Discovering new solutions to old problems is very liberating and exciting.

So the second thing we will emphasize in this handbook is understanding what makes the models tick. We will present several techniques for network analysis, and try to extract the basic principles which are responsible for the networks' behavior. Network analysis is not easy, and the tools we suggest here only scratch the surface of what one would like to do. The problem is not unlike the one we would encounter if we actually did get to open up our subjects' heads: It's pretty complicated in there. The difference, of course, is that our models are vastly more tolerant of being poked and probed and lesioned than are our human subjects. And they recover faster!

Do you need to know how to program in order to do simulations? No. Ultimately it's helpful, but a great deal can be done without actually having to do any programming, and no programming at all is required to use this handbook. The handbook is organized around a set of simulations which can be run using the software which is provided on the enclosed floppy disks. This software includes a neural network simulator called **tlearn**. Having a simulator allows you to experiment with different network architectures and training regimes without having to actually program the networks yourself (e.g., by writing software in C or Fortran). So if you do not know how to program, don't worry. The simulator simply needs to be told what the network's architecture is, what the training data are, and a few other parameters such as learning rate, etc. The simulator creates the network for you, does the training, and then reports the results. Other tools are included which allow you to examine the network's solutions. Thus, while you might find it eventually useful to be able to program yourself (for example, to create new stimuli), you will not need to do any programming to run the simulations described in this book. You will need to be able to use a simple text editor; whichever one you use is up to you.

The software itself exists in several versions, and will run on Macintoshes, PCs, Unix workstations, and most other machines which have ANSI standard C compilers. Precompiled binaries exist for several common platforms, but sources are provided in case you wish to modify and recompile the simulator yourself. Chapter 3 provides an overview of the simulator; installation procedures and reference manuals are supplied in Appendices A and B. When we designed our first few exercises, we kept in mind that most readers would need explicit help in using the simulator. These exercises should help ease the reader into the process of running a simulation. We do not guarantee that the software provided with this book is free of bugs. In fact, we guarantee that if you try hard enough you will find some situations where **tlearn** will break. Please tell us about any software problems you experience by emailing us at innate@crl.ucsd.edu. We may not respond to your queries immediately but we will fix the bugs as time permits. You can obtain the latest versions of **tlearn** via anonymous ftp at crl.ucsd.edu in the pub directory or on the world wide web at http://crl.ucsd.edu/innate or via anonymous ftp at ftp.psych.ox.ac.uk also in the pub directory.

A very major source of support—intellectual as well as monetary—for this project came from the MacArthur Foundation through their funding of the Neural Modeling Training Program for Developmentalists which was run at the Center for Research in Language at

UCSD from 1988 to 1993. We believed, and MacArthur was willing to gamble, that the connectionist paradigm has great potential for developmentalists. Through the training program we were able to bring to CRL more than a dozen developmentalists, junior researchers as well as senior people, for periods ranging from a week to several months. The training program worked in two directions. Trainees learned how to model; we got to pick their brains. It has been a stimulating and rewarding experience and we are enormously grateful to our trainees and to MacArthur for the willingness to embark on this venture with us. The far-sighted approach of Bob Emde, Mark Appelbaum, Kathryn Barnard, Marshall Haith, Jerry Kagan, Marion Radke-Yarrow, and Arnold Sameroff was critical in this effort. Without their support—both tangible as well as intellectual—we could not have written this book.

We also thank our trainees: Dick Aslin, Alain Content, Judith Goodman, Marshall Haith, Roy Higginson, Claes von Hofsten, Jean Mandler, Michael Maratsos, Bruce Pennington, Elena Pizzuto, Rob Roberts, Jim Russell, Richard Schwartz, Joan Stiles, David Swinney, and Richard Wagner. Neither this nor the companion volume (*Rethinking Innateness*) could have been written without their excitement, inspiration, and practical advice. And we are very grateful to those who served as trainers in this program: Arshavir Blackwell, Mary Hare, Cathy Harris, and Virginia Marchman.

In addition, the material contained in the book has been used by participants in several of the Oxford Connectionist Modeling Summer Schools sponsored by the Oxford McDonnell-Pew Centre for Cognitive Neuroscience. We should like to thank the Summer School participants for their feedback on earlier versions of the exercises and the McDonnell-Pew Foundation for their generous support of this annual meeting. The demonstrators on the summer school—Todd Bailey, Neil Forrester, Patrick Juola, Denis Mareschal, Ramin Nakisa, Graham Schafer and Michael Thomas—helped us spot many errors in earlier drafts and helped us to realize that what seemed obvious was often a mystery.

Last, and certainly not least, we should like to express our gratitude to John Kendall and Steven Young for their heroic efforts in responding to our requests to continually update the **tlearn** interface, add new utilities and then change everything back again. Without their patience, determination and fine programming skills, this book would simply not exist.

How to use this book

This book is designed as a companion volume to *Rethinking Innateness: A connectionist perspective on development.* That book is our attempt to sketch out what we hope may serve as a framework for a theory of development. The book necessarily omits some details, particularly regarding finer details of simulations. One purpose of this book is to provide a way for the interested reader to pursue in depth some of the issues which we raise in the first volume.

Our larger ambition is that we will have persuaded some readers that this is a useful methodology which can serve them in their own research. We hope that this volume will therefore provide those readers with tools which they can use on their own. There are other simulators, of course, and architectures which may require more special-purpose software. This simulator, however, has served us well and is used by us not only for pedagogical purposes, but in our own research. We hope it will serve you as well!

KP & JLE

Oxford
La Jolla
1997

CHAPTER 1   *Introduction and overview*

Our goal in this book is to illustrate, in concrete ways that you will be able to replicate on your own, properties of connectionist models which we believe are particularly relevant to developmental issues. Our emphasis on the principles and functional characteristics of these models is what sets this book apart from many of the other excellent introductions to neural networks (some of which the reader may wish to consult to get a broader view of architectures and techniques not covered in this volume).

In Chapters 3 through 12 we explore a set of simulations which focus on various aspects of connectionist models. However, we are aware that our readers will vary widely with regard to the knowledge and experience they bring with them. Before leaping into the simulations, therefore, there are several things we think it will be useful to do. The first three chapters therefore provide an overview of some of the technical aspects of doing simulations. In this chapter we introduce some of the terminological and notational conventions which will be used in this book, and provide a brief overview of network dynamics and learning. Our intent is modest here; we want to give the reader enough of an understanding of network mechanics so that he or she will understand what actions are being done by the simulator that is used in the subsequent exercises. Our goal in Chapter 2 is to make explicit the assumptions which underlie the simulation methodology we will be using. It is easy to do simulations; it's not as easy to do them well and to good purpose! In Chapter 3 we describe the software which will be used in this book. These first three chapters thus contain introductory material, some of which the experienced reader might wish to skip (although we urge that it at least be skimmed to ensure nothing vital is missed).

Nodes and connections

Neural networks are actually quite simple. They are made up of a few basic building blocks: *nodes* and *connections*. Figure 1.1 shows sev-

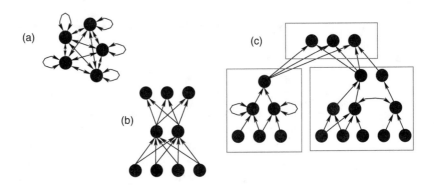

FIGURE 1.1   Various types of connectionist architectures. (a) A fully recurrent network; (b) a three-layer feedforward network; (c) a complex network consisting of several modules. Arrows indicate direction of flow of excitation or inhibition.

eral sample networks, where nodes are shown as filled circles and connections as lines between them.

Nodes are simple processing units. They are often likened to neurons. Like neurons, they receive inputs from other sources. These inputs may be excitatory or inhibitory. In the case of the neuron, excitatory inputs tend to increase the neuron's rate of firing, while inhibitory inputs decrease the firing rate. This notion of firing rate is captured in nodes by giving them a real-valued number which is called their *activation*. (We might think of higher activation values as corresponding to greater firing rates, and lower activation values to lower firing rates.)

The input to a given node comes either from other nodes or from some external source, and travels along connection lines. In most connectionist models, it is useful to allow connections between different nodes to have different potency, or connection strengths. The strength of a connection may also be represented by a real-valued number, and is usually called the *connection weight*. The input which flows from one node to another is multiplied by the connection weight. If the connection weight from one node to another is a negative number, then the input from the first to the second node may be thought of as being inhibitory; if positive, it is excitatory.

If we looked in more detail at a node, we might wish to represent it as in Figure 1.2. This shows the node as a circle, with input connec-

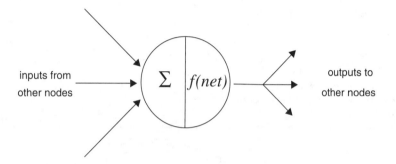

FIGURE 1.2   Detailed look at a single node. Inputs to the node are typically summed (indicated by the symbol Σ on the left); the net input is then passed through an *activation function* (shown as *f(net)*) which yields the node's activation. This value is then sent on to other nodes.

tions feeding into it, and output connections leading from it. Each input line or connection represents the flow of activity perhaps from some other neuron or from some external source (such as light falling on some photosensitive retinal cell).

For most of the nodes that you will meet in this book, the cell body performs two operations. The first is the simple adding together of the net inputs to the unit. Each input (from different nodes) is itself a number, which can be calculated by multiplying the activation value of the sending node by the weight on the connection from the sending to receiving node (note that connections between nodes may be asymmetric). If we use the letter i to index the receiving node, a_j to index the activation of those nodes which send to node i, and w_{ij} to refer to the weights on the connections from nodes j to node i, then we may calculate the net input to node i as

$$\text{net}_i = \sum_j w_{ij} a_j \qquad \textbf{(EQ 1.1)}$$

A word on notation

What does Σ mean?

When you see this symbol (called "sigma") it means that something is going to be added. Here we use it by itself to indicate that all the inputs will be summed together. Another example might be

$$\sum_i a_i$$

This means that we have some number of a's to be summed. We use i as a counter, beginning with i equal to 0 (by convention). We sum the first a (a_0), the second, (a_1), etc., till we have counted through all "i" of them.

What does $f\,(net)$ mean?

This is read "f of net." It means we have some operation or set of operations which we want to carry out on the quantity contained in the variable named "net" (which we use to denote the net input to a node). We call these operations a function, and say that we are applying the function "f" to the quantity "net." (Note that so far, we have left unspecified just what that function is.)

What a node actually *does* with that net input is another matter. In the simplest case the node's activation is the same as its input. In this case the activation function (*f(net)*) is just the identity function. But one can easily imagine cases where the activation of a node (its output) might require a certain amount of "juice" before it actually starts to

Unpacking Equation 1.1

This equation tells us how to calculate the total input coming into some node. We call that node i so that the procedure can be general. Let's assume here we are dealing with node 5, so i equals 5. We have already said that the Σ means to add some things together; the subscript under the sigma (j) tells us how many things need to be added. The things to be added are indicated by the letters that follow—the w_{ij} and the a_j. By convention, two adjacent variables mean the numbers they represent are to be multiplied first. So, how do we calculate all of this?

We begin by setting our counter (j) to 0 (again, by convention). That means a_j is a_0, or the activation of the 0th node (whatever it happens to be). w_{ij} becomes the $w_{5,0}$ or the weight going to node 5 from node 0. We multiply these two numbers together and save the result. Then we set the counter j to 1, and calculate the product of a_1 (the activation of node 1) times $w_{5,1}$ (the weight to node 5 from node 1). We save that result. We continue till we have gone through all of the j's. Finally, we add them up (the sigma). This operation is often called a "sum of products."

fire. This is in fact typical of real neurons: In order to begin firing, the input must exceed a certain threshold.

In many neural networks, the activation function is nonlinear function of the input, resembling a sigmoid. In the networks we will be using here, nodes' activations are given by the logistic function shown in

$$a_i = \frac{1}{1 + e^{-net_i}}$$ (EQ 1.2)

(where a_i refers to the activation (output) of node$_i$, net_i is the net activation flowing into node$_i$, and e is the exponential). This equation tells us what the output of a node will be, for any given net input. If we graph this relationship, as we have done in Figure 1.3, we get a

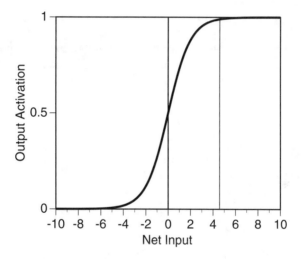

FIGURE 1.3   The sigmoid activation function often used for nodes in neural networks.

better idea of how a node's output is related to its input.

We see that over a wide range of inputs (roughly, inputs greater than 4.0, or less than -4.0), such nodes exhibit an all-or-nothing response—they are either fully "on" (output their maximum values of 1.0) or "off" (output their minimum values of 0.0). Within the range of -4.0 to 4.0, on the other hand, the nodes show a greater sensitivity and their output is capable of making fine discriminations between different inputs. ***This nonlinear response lies at the heart of much of what makes such networks interesting.***

A concrete example

Although the dynamics of node activations are fairly straightforward, it is easy to be confused between the *input* which a node receives, and its *output*. Calculating these quantities is one of the things a simulator does, but these can also be calculated by hand and it is useful to do this a few times to be sure you understand what is going on.

To place things in context, let us first assemble a simple network. A neural network consists of a collection of nodes of the sort that we discussed in the previous section. When we talk about the *architecture* of a network we are referring to the particular way in which that network is assembled, or its pattern of connectivity. There are many types of architectures, and we shall consider a number of them in this book.

A very common architecture is one in which nodes are connected to each other in a layered fashion. For example, consider the neural network depicted in Figure 1.4. This network consists of four nodes

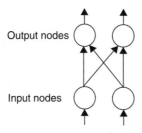

FIGURE 1.4   A two-layered feedforward network.

organized into two layers: an input layer and an output layer. Within the input layer, all the nodes have connections which project to the output layer. There are no connections between nodes within a layer (no *intra*-level connections). Furthermore, in this architecture the nodes do not possess *recurrent* connections, i.e., they do not have connections which project back to themselves or to lower levels. Thus, in this network, the flow of activity is in one direction only, from the input layer to the output layer. We call these types of networks "feedforward networks." In contrast, "recurrent networks" may possess both intra- and inter-level connections as well as feedback connections from one level to an earlier level.

Notice that the input nodes in Figure 1.4 have only a single connection projecting into them. Similarly, the output nodes have only a single connection projecting out from them. Again, this portrayal is a gross simplification in comparison to biological neural networks. Real neural networks are likely to receive inputs from multiple sources and send outputs to multiple destinations. Of course, there will be some biological neurons that receive only a single input. For example, retinal photoreceptors might be thought as neurons with just a single input—in this case, the light source that fires the neuron. More generally, though, it is appropriate to think of the single input to an input neuron in Figure 1.4 as summarizing the input from multiple sources, and the single output from an output neuron as summarizing the output to multiple destinations.

We can now begin to consider just how the neural network performs its task. First, let's assume that each input node has a certain level of activity associated with it. Our goal is to determine how the activity of the input neurons influence the output nodes. To simplify the explanation, we shall consider the process from the point of view of just one output unit, the left-hand output in Figure 1.4. This is highlighted in Figure 1.5. We refer the two input nodes as $node_0$ and

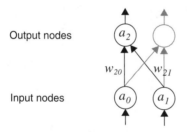

FIGURE 1.5   The activation of the left-hand output unit from Figure 1.4.

$node_1$, and to the two output nodes as $node_2$ and $node_3$. The activation values of the input nodes are denoted a_0 and a_1, respectively. Our goal is to calculate the activation of the left-most output node, a_2.

From Figure 1.2 we see that one of the computations that the neuron performs is to calculate its net input from other neurons. The output neuron in Figure 1.5 receives input from two input neurons, namely a_0 and a_1. These two input neurons communicate with the output neurons via independent connections. We also said earlier that exactly how much input was received along a given connection depended on the activation values of the sending units (in this example, a_0 and a_1), but also the weights on the connections. These weights serve as multipliers. In Figure 1.5 we have denoted the weight from input $node_0$ to output $node_2$ with the symbol w_{20}, using the convention that a weight labeled w_{ij} refers to the connection *to* $node_i$ *from* $node_j$. Note that since activity flows in only one direction along the connections, the value of the weight w_{20} is not the same as w_{02}. In fact, the connection w_{02} does not exist in the network depicted in Figure 1.5.

In the example above, the only inputs to $node_2$ come from the two input nodes. Each input is the product of the activation of the sender unit times the weight; the total input to $node_2$ is simply given by the sum of these two products, i.e., $w_{20}a_0 + w_{21}a_1$.

Exercise 1.1 [a]

To make the example concrete, assume our network has the weights shown, and the input nodes have the activations shown.

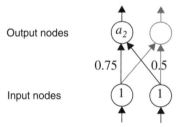

1. What will be the *input* which is received by $node_2$?

The net input by itself does not determine the activity of the output node. We also need to know the activation function of the node. Let us assume our nodes have activation functions as given in Equation 1.2 (and shown graphically in Figure 1.3). In the table below we give sample inputs and the activations they produce, assuming a logistic activation function.

Exercise 1.1 [a]

INPUT	ACTIVATION
-2.00	0.119
-1.75	0.148
-1.50	0.182
-1.25	0.223
-1.00	0.269
-0.75	0.321
-0.50	0.378
-0.25	0.438
0.00	0.500
0.25	0.562
0.50	0.622
0.75	0.679
1.00	0.731
1.25	0.777
1.50	0.818
1.75	0.852
2.00	0.881

2. What will be the *activation* of $node_2$, assuming the input you just calculated?

a. Answers to exercises are given at the end of each chapter.

In many networks, it is also useful to allow nodes to have what amounts to a default activation. Note that in the absence of any input (which means an input of 0.0), our nodes will have an output of 0.5 (see Exercise 1.1). Suppose we want a node to be "off"—have an output of 0.0—in the absence of input. Or we might wish its default state to be on.

We can accomplish this by adding one additional node to our network. This node receives no inputs, but is always fully activated and outputs a 1.0. The node can be connected to whatever other nodes in the network we wish; we often connect this node to all nodes except the input nodes. Finally, we allow the weights on the connections from this node to its receiving nodes to be different.

This effectively guarantees that all the receiving nodes will have some input, even if all the other nodes are off. Since the extra node's output is always 1.0, the input it sends to any other node is just $1.0 \times w_{ij}$—or the value of the weight itself.

Because of what it does, this extra node is called the *bias* node (only one is needed per network). What it does is similar to giving each node a variable threshold. A large negative bias means that the node will be off (have activations close to 0.0) unless it receives sufficient positive input from other sources to compensate. Conversely, if the bias is very positive, then the receiving node will by default be on and will require negative input from other nodes to turn it off. Allowing individual nodes to have different defaults turns out to be very useful.

Learning

So far we have discussed simple networks that have been pre-wired. In Exercise 1.1 we gave as an example a network whose weights were determined by us. For some other problem, we might wish the network to learn what those weights should be.

In this book we will be using a learning algorithm called "backpropagation of error" (Rumelhart, Hinton, & Williams, 1986; see also Le Cun, 1985; Werbos, 1974). Backpropagation is also referred to as the 'generalized delta rule'. (This algorithm is described fully in the paper by Rumelhart et al. (1986) and the reader is urged to consult that paper for a more detailed explanation.)

The basic strategy employed by "backprop" is to begin with a network which has been assigned initial weights drawn at random, usually from a uniform distribution with a mean of 0.0 and some user-defined upper and lower bounds (frequently ±1.0). The user also has a set of training data, which come in the form of *input/output* pairs. The goal of training is to learn a single set of weights such that any input pattern will produce the correct output pattern. Often it is also desired that those weights will allow the network to generalize to novel data not encountered during training.

The training regime involves several steps. First, an input/output pattern is selected, usually at random. The input pattern is used to activate the network, and activation values for output nodes are calculated. (Note that in the example in Figure 1.4 our network has only input nodes and output nodes. We could just as easily have additional nodes between these two layers, and in fact there are good reasons to

What is $f'(\text{net}_{ip})$?

$f'(net_{ip})$ *(pronounced "f prime of net$_{ip}$") is the first derivative of the node's activation function. This is just the slope of the activation function. The activation function, the sigmoid, is defined mathematically in* **(EQ 1.2)** *and depicted graphically in Figure 1.3. Notice that the slope is steepest around the middle of the function (where the net input is close to zero). In fact, the slope of the sigmoid activation function is given precisely by the expression $o_{ip}(1 - o_{ip})$. This is plotted in Figure 1.6.*

The error term δ_{ip} is just the product of the actual error on the output node and the derivative of the node's activation function. For large values of net input to the node (both positive and negative) the derivative is small. Consequently, δ_{ip} will be small. Net input to a node tends to be large when the connections feeding into the node are strong. Conversely, weak connections tend to yield a small input to a node. With small values of net input, the derivative of the activation function is large (see Figure 1.6) and δ_{ip} can be large—provided the output error is large.

wish to have such "hidden nodes"; see the companion volume *Rethinking Innateness*, Chapter 1 and Chapter 3, this volume.)

Because the weights of the network have been chosen at random, the outputs that are generated at the outset of training will typically not be those that go with the input pattern we have chosen; the outputs are more likely to be garbage than anything else. In the second step of training, we compare the network's output with the desired output (which we call the *teacher* pattern). These two patterns are compared on a node-by-node basis so that for each output node we can calculate its error. This error is simply the difference in value between the target for node$_i$ on training pattern p (we will call this target t_{ip}) and the actual output for that node on that pattern (o_{ip}), multiplied by the derivative of the output node's activation function given its input. We'll call that error δ_{ip} (δ is pronounced "delta"):

$$\delta_{ip} = (t_{ip} - o_{ip})f'(net_{ip}) = (t_{ip} - o_{ip})o_{ip}(1 - o_{ip}) \qquad \textbf{(EQ 1.3)}$$

So the problem now is how to apportion credit or blame to each of the connections in the network. We know, for each output node, how far off the target value it is. What we need to do is to adjust the weights on the connections which feed into it in such a way as to reduce that error. That is, we want to change the weight on the connections from every node$_j$ coming into our current node$_i$ in such a way that we will reduce the error on this pattern. This change in weight is calculated as:

$$\Delta w_{ij} = -\eta \frac{\partial E_p}{\partial w_{ij}}$$

(EQ 1.4)

What does $\frac{\partial E_p}{\partial w_{ij}}$ mean?

*This is what is called a **partial derivative**. What it expresses is actually very straightforward and can be understood intuitively without knowing calculus. Basically, this term measures how the quantity on the top changes when the quantity on the bottom is changed. In this particular case, we want to know how the error (E) is affected by changing the weights (w).*

If we knew this, then we would know how to change the weight (the Δ symbol—also pronounced "delta"—on the left of Equation 1.4 means "change") in order to decrease the error, where error will mean the discrepancy between what the network is outputting, compared with what we want it to be outputting.

That is, we want to know how changes in error are related to changes in weights. (The η —pronounced "eta"—is known as the learning rate, and is a small constant. Since our goal is to find a set of weights which will work for *all* input/output patterns, we should be cautious in changing the weights too much on any given pattern.)

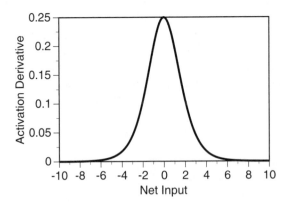

FIGURE 1.6   The derivative of the activation function

Of course, the real question is how we compute the expression on the right, practically speaking. It turns out that this quantity can be calculated as:

$$\Delta w_{ij} = \eta \delta_{ip} o_j$$ (EQ 1.5)

This is often called the "delta rule." We won't explain the math behind the derivation here; if you are interested, consult Rumelhart et al., 1986; or Hertz, Krøgh & Palmer, 1991.

We do three things with this equation. First, we make our changes small, so η is often set to some value less than 1.0 (e.g., 0.1 or 0.3). We do this because, if we are updating weights after every pattern, we don't want to have changes be too drastic. There are other patterns yet to be encountered, and we wish to proceed cautiously so that we can find a set of weights which will work for all the patterns, not just the current one. Second, the change in weight depends on the error we have for this unit, δ_{ip}, as calculated in Equation 1.3. Finally, we also take into account the output we have received from the sending node, o_j. This makes sense because the node's error is related to how much (mis)information it has received from another node; if the other node is highly active and has contributed a great deal to our current activation, then it bears a large share of the responsibility for our error.

We proceed in this manner, calculating errors on all output nodes, and weight changes on the connections coming into them (but we do not yet actually make the changes). Then we move down to the hidden layer(s) (if there are any). We use the same equation, Equation 1.5, for changing weights that lead into the hidden units from below. However, we cannot use Equation 1.3 to compute the hidden nodes' errors, since there is no given target against which they can be compared. Instead, we make the hidden nodes "inherit" the errors of all the nodes they have activated, using the same principle of credit/blame. If the nodes activated by a hidden node have large errors, then the hidden unit shares responsibility. So we calculate its error by simply summing up the errors of the nodes which it activates (multiplied by the weight between the nodes, since obviously if the weight is very small the hidden node has much less responsibility). This procedure is summarized in Equation 1.6 where the subscript i indicates the hidden node, p indicates the current pattern and k indexes the output node feeding error back to the hidden node:

$$\delta_{ip} = f'(net_{ip})\sum_k \delta_{kp} w_{ki} \qquad \text{(EQ 1.6)}$$

(As in Equation 1.3, the derivative of the hidden unit's activation function is also multiplied in.)

This procedure continues iteratively down through the network, hence the name "backpropagation of error." When we get to the layer above the input layer (inputs have no incoming weights), we take the third and final step of actually imposing the weight changes.

Exercise 1.2

---

1. Why do we wait until we have calculated all the δ s before making the weight changes, rather than change weights as we go down the network?

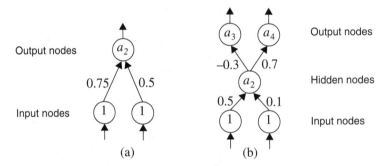

(a)   (b)

2. Imagine we are training the single-layered network shown above on the left. The network is shown with a set of activation values for the input nodes and weights connecting the input nodes to the output node. Assume that the output node has a sigmoid activation function (see Exercise 1.1), that the desired output is 1.0 and that the learning rate $\eta = 0.1$. Calculate the changes that will be made to the two weights in the network. *Hint: You will also need to know the value of the derivative in Equation 1.3. These are tabulated for different activation values below. Don't forget to calculate net inputs to determine the activation value from the table in Exercise 1.1.*

Exercise 1.2

ACTIVATION	DERIVATIVE
0.0	0.00
0.1	0.09
0.2	0.16
0.3	0.21
0.4	0.24
0.5	0.25
0.6	0.24
0.7	0.21
0.8	0.16
0.9	0.09
1.0	0.00

3. Repeat the calculations for the multi-layered network with hidden nodes using the same learning rate parameter and target activations of 1.0 on both output nodes. *Hint: Calculate δ s for the output nodes in just the same way as you did in Exercise 1.2.1. However, you will also need to consult Equation 1.6 to determine δ s for the hidden node.*

4. What training conditions promote maximum weight changes in a network? *Hint: Consult Equation 1.3 and Equation 1.5 before attempting to answer this question.*

Answers to exercises

Exercise 1.1

1. If the activation values of the two input nodes were 1.0 and 1.0, respectively and the connection weights were 0.75 and 0.5 then the net input to $node_2$ would simply be the sum of the products of each weight by the input coming in along that weight:

    $$netinput_2 = (1.0 \times 0.75) + (1.0 \times 0.5) = 1.25.$$

2. If $node_2$ receives this input, we can read off its corresponding activation by consulting the activation table in Exercise 1.1. We see that an input of 1.25 leads to an activation of 0.777. Note that the activation function scales inputs into the range from 0.0 to 1.0. When the net input is 0.0, the node's output is exactly in the midrange of its possible activation range: 0.5. Positive inputs result in activations that are greater than 0.5; negative inputs result in activations that are less than 0.5.

Exercise 1.2

1. After we compute the weight changes for a layer, we hold off actually changing weights until at least after we have calculated the errors on the nodes *below* us. This is because the errors for those nodes are calculated using Equation 1.6. In that equation, the nodes below a layer inherit some part of the error of the layer above; how much error is based on the error itself, multiplied by the weights from the lower to the upper layers. So we do not want to change those weights until after we have apportioned blame, using the weights that were in effect at the time the output pattern was generated.

2. As we saw in Exercise 1.1 the activation of the output node is 0.77. This means that the discrepancy between the actual output and the desired output $t - o = 1.0 - 0.77 = 0.23$. The calculation of δ from Equation 1.3 requires we determine the derivative $f'(net)$ for an activation of 0.77. From Exercise 1.2 we see that an activation value of 0.77 yields a derivative value of approximately 0.16. Substituting these values in Equation 1.3 we get:

$$\delta_2 = (t-o)f'(\text{net}_2) = 0.23 \times 0.16 = 0.037$$

We are now in a position to calculate the individual weight changes from Equation 1.5:

$$\Delta w_{20} = \eta \delta_2 o_0 = 0.1 \times 0.037 \times 1.0 = 0.0037$$

Recall that the constant η defines the learning rate and o_o stands for the activation of the input node 0. Similarly, the weight change for the connections from input node 1 to the output node is given by:

$$\Delta w_{21} = \eta \delta_2 o_1 = 0.1 \times 0.037 \times 1.0 = 0.0037$$

In this example, the changes to each of the weights are identical since the input nodes have the same activation. Note that the overall effect of this training trial is to change the connections only slightly. The process would have to be repeated many times before the error is reduced to zero. Keeping the weight changes small helps prevent the network making adjustments that might not be suitable for other input/output patterns.

3. First calculate the activations of all the receiving nodes:

$$a_2 = \text{logistic}((1.0 \times 0.5) + (1.0 \times 0.1)) \approx 0.65$$
$$a_3 = \text{logistic}(0.65 \times -0.3) \approx 0.45$$
$$a_4 = \text{logistic}(0.65 \times 0.7) \approx 0.6$$

Second calculate $\delta_{3,4}$ for the output nodes:

$$\delta_3 = (t_3 - o_3)f'(\text{net}_3) \approx (1.0 - 0.45) \times 0.24 = 0.13$$
$$\delta_4 = (t_4 - o_4)f'(\text{net}_4) \approx (1.0 - 0.6) \times 0.24 = 0.1$$

Now calculate δ_2 for the hidden unit using Equation 1.6:

$$\delta_2 = f'(\text{net}_2) \sum_{k \in 3,4} \delta_k w_{k2}$$
$$= 0.23((0.13 \times -0.3) + (0.1 \times 0.7))$$
$$= 0.007$$

We can now determine all the weight changes using Equation 1.5:

$$\Delta w_{20} = \eta \delta_2 o_0 = 0.1 \times 0.007 \times 1.0 = 0.0007$$
$$\Delta w_{21} = \eta \delta_2 o_1 = 0.1 \times 0.007 \times 1.0 = 0.0007$$
$$\Delta w_{32} = \eta \delta_3 o_2 = 0.1 \times 0.13 \times 0.65 = 0.008$$
$$\Delta w_{42} = \eta \delta_4 o_2 = 0.1 \times 0.1 \times 0.65 = 0.007$$

Note how all the weight changes are made *after* the δs for output and hidden nodes have been computed. The learning rate $\eta = 0.1$ scales the size of the weight changes and thereby helps to ensure the network is not too influenced by errors for individual patterns.

4. There are several factors that determine how much the weights are changed on any learning trial. These are summarized in Equation 1.5. The learning rate influences the proportion of the error term δ that is used to change the weights. So large learning rates will tend to lead to large weight changes. The error term δ itself is defined as the product of the actual output error and the derivative of the sigmoid activation functions (see Equation 1.3). Large errors will also tend to lead to large weight changes. However, the effect of the error is modified by the derivative of the sigmoid. We saw in Figure 1.6 that the derivative is largest when the input to the node is small. Small inputs tend to go together with small weights. Networks tend to have small weights at the beginning of training. Hence, networks are most sensitive to learning early in their lives. As the network gets older, its weights get bigger leading to large net input values. Large net inputs give small derivatives, leading to small weight changes. Older networks find it more difficult to learn!

## CHAPTER 2

The methodology of simulations

Why build models?

The idea of building formal models to test a theory of behavior is not a new one, nor original with connectionists. Models in general play an important role in the behavioral sciences; one has only to look through a text in mathematical psychology, in economics, or in sociology to see this. In fact, a model is simply a detailed theory. Actual computer simulations of models are perhaps not so common, but they play a special role in connectionism.

There are several reasons why computer simulations are so useful. First, there is the matter of explicitness. Constructing a model of a theory or an account, and then implementing it as a computer program, requires a level of detail which is not required by mere verbal description of the behavior. These details may turn out to be crucial for testing the model. Even the process of converting the model to a program can be useful, because it encourages us to clarify our thinking and consider aspects of the problem we might have skipped over if we were simply describing the theory in words.

Second, there is the fact that it is often difficult to predict what the consequences of a model are, especially if it is at all complicated. There may be interactions between different parts of our model which we cannot work out in our heads. Connectionist models have the additional problem that, although in some ways they are very simple, they involve processing elements which are nonlinear. While linear systems can sometimes be analyzed in advance, nonlinear systems frequently need to be simulated and then studied empirically. This is common practice among physicists who study nonlinear dynamics, for example; much of their work is empirical. Frequently the only way to see how a model will behave is to run it and see.

Third, we find that the unexpected behaviors exhibited by the simulations can suggest new experiments we can run with humans. We often construct our models to fit some behavioral data we are interested in understanding. But once we've got the simulation up and running, we can see how it will respond in novel situations. In many cases, these responses are unanticipated and let us make predictions about how humans would respond in these untested situations. We can run these experiments, and if our models are good—and we are lucky!—the new data may bear out the predictions. If they don't, that's useful information too, because it lets us know there is something wrong in the model.

Fourth, there may be practical reasons why it is difficult to test a theory in the real world. For example, if we have hypotheses about the effects of brain damage on human subjects, our "experiments" are limited to the tragedies of nature. The paucity of available data may make it difficult to see global patterns. With a model, we can systematically vary parameters of interest through their full range of possible values. This often reveals patterns which we might have otherwise missed. We can then go search for corresponding instances in the real world, armed with our model's predictions. Models can thus help stimulate new empirical work, as well as describe existing data.

Finally, and perhaps most important, simulations play an important role in helping us understand *why* a behavior might occur. Sometimes it is enough to just build a simulation and have it do what we want. But in most cases, what we're after is an explanation. When we work with human subjects, we build hypotheses, based on observed behaviors, about why they act as they do. We try to be creative and think of clever ways to test our hypotheses. We almost never can get inside the subject to verify whether our theories are accurate. Our computer simulations, on the other hand, are open for direct inspection. We need not be content with simply making theories about their behavior. We can actually open them up and peer at the innards. Doing this is not always easy, but we think it is one of the most important reasons to do simulations. For this reason, we will spend a lot of time in this book talking about network analysis and about understanding why networks work the way they do. We will also focus on trying to understand the basic principles which underlie the simulations so that we can go beyond the specifics of any given simulation.

We should add a word of caution here. Computer models play a similar role in behavioral analysis as animal models play in medicine and the biological sciences. Both sorts of models are metaphors. As such, they are only as good as their resemblance to the real system whose behavior they claim to model. Some animal models are very useful for understanding mechanisms of human physiology, because the animals are known to have close anatomical or physiological similarities to humans. Other animals make poor models (we wouldn't want to study human vision using the fruitfly, for example). The same holds true for computer models.

This is one reason why the question of a model's plausibility is so important. We need to worry not only about the neural plausibility of the model, but also it's psychological plausibility. We want our models to be learning behaviors which we think resemble those of humans. And we want the models to be given the same kind of information which we think is plausibly available to humans as well. This is not always an easy thing to know.

There is never any guarantee that the model is an accurate mirror of the human. The most we can say is that there is such a close correspondence between the behaviors, and between the kinds of constraints which are built into the model and which operate in humans, that we believe the model captures the essence of what is going on in the human.

The exciting thing is that the study of these artificial systems can also liberate us from biases and preconceptions which we might not even be aware of. We approach our study of behavior with a rich legacy of theories. Sometimes this legacy blinds us to ways of seeing phenomena which might lead to new analyses. Connectionist models often exhibit behaviors which are eerily like those of humans. We are surprised when we look inside the model to see what it is really doing, and discover that the model's solution to the problem may be very different from what we assumed was the obvious and only reasonable solution. Sometimes the model's solution is much simpler than our own theories; sometimes it is more complicated. But often, the solution is different, and can generate new hypotheses and insights.

Simulations as experiments

It's easy to do simulations, but hard to do them well. The ready availability of off-the-shelf neural network simulators makes it comparatively easy to train networks to do lots of things. Much of the initial flurry of excitement and interest in neural network learning was brought about by the discovery that networks could be successfully trained on a wide range of tasks. During the early period of connectionist research, exploring the many things that networks could do was intrinsically interesting.

But in the long haul, novelty wears off. We begin to find, too, that it is actually not all that easy to train networks on many tasks. And even when a network is trained successfully, we begin to ask ourselves, "So what? What have we learned or demonstrated that we did not know before?"

In fact, in our view, running a good simulation is very much like running any good experiment. We begin with a *problem* or goal that is clearly articulated. We have a well-defined *hypothesis*, a *design for testing* the hypothesis, and a plan for how to *evaluate the result*. And all of this is done *prior* to ever getting close to the computer.

The hypothesis typically arises from issues which are current in the literature. Simulations are not generated out of the blue. The nature of the hypothesis may vary. In some cases we may wish to test a prediction that has been made. Or we might wish to see whether an observed behavior might be generated by another mechanism (for example, as a way of testing the claim that only a certain mechanism is able to produce the behavior). Or we might have a theory of how some behavior is accomplished, and the simulation lets us test our model. Only rarely do we indulge in "fishing expeditions" (in which we start with no clearly defined goal); when we do, we must be prepared to come home empty-handed.

The hypothesis and goals of the simulation must be formulated very explicitly before we begin to think about the design. This is crucial, because when the simulation is done, we will need to know how to evaluate the results. What tests we run to evaluate the simulation will depend on what our initial questions were. Furthermore, our ability to run these tests will be limited by the design of the simulation. So the design itself is highly constrained by the hypothesis.

Such considerations are of course true of any experiment. We are simply emphasizing that these methodological constraints are equally true of simulations. However, there are some ways in which the correspondences between running simulations and running experiments with human subjects might not be as obviously similar. These largely have to do with design and evaluation issues.

The design of a simulation involves several things which may be unfamiliar to some. These include the notion of a *task, stimulus representations*, and *network architecture*. Ultimately, how to find the right task, representation, and architecture are best illustrated by example. That is one of the purposes of the simulation exercises which follow in the remaining chapters. But we think it is useful to begin with some explicit discussion of these issues to help focus your attention later as you work through the exercises.

The task

When we train a network, our intent is to get it to produce some behavior. The *task* is simply the behavior we are training the network to do. For example, we might teach a network to associate the present tense form of a verb with its past tense form; to decide whether or not a beam with two weights placed at opposite ends will balance; to maintain a running count of whether a digit sequence sums to an odd or even number; or to produce the phonological output corresponding to written text.

We need to be a bit more specific, however. What does it mean to teach a network to read text, for example?

In the networks we will be dealing with in this book, a task is defined very precisely as learning to produce the correct output for a given input. (There are other ways of defining tasks, but this is the definition that works best for the class of networks we will use.) This definition presupposes that there is then a set of input stimuli; and paired with each input is the correct output. This set of input/output pairs is often called the "training environment."

There are several important implications of this which we need to make explicit. First, we will have to be able to conceptualize behaviors in terms of inputs and outputs. Sometimes it will be easy to do this, but other times we may have to adopt a more abstract notion of what constitutes an input and what constitutes an output. We'll give a simple example. Suppose we wish to teach a network to associate two forms of a verb. Neither form properly constitutes an input, in the sense that it is a stimulus which elicits a behavior. However, we might nonetheless train the network to take one form as input and produce the paired associate as output. This gives us the input/output relationship that is required for training, but conceptually we might interpret the task in more abstract terms as one of *associating* two forms.

Secondly, note that we are teaching the network a task by example, not by explicit rule. If we are successful in our training, the network will learn the underlying relationship between the input and output by induction. This is an appealing model for those situations where we believe that children (and others) learn by example. However, it is extremely important not to assume that the network has in fact learned the generalization which *we* assume underlies the behavior. The network may have found some other generalization which captures the data it has been exposed to equally well, but is an easier generalization to extract, given the limited data to which the network has been exposed. (What, precisely, is meant by "easier" is an important and fascinating question which is not fully understood.) This is where our testing and evaluation become important, because we need ways to probe and understand the content of the network's knowledge.

Finally, a concomitant of this style of inductive learning is that the nature of the training data becomes extremely important for learning. In general, networks are data grubs: They cannot get too much information! The more data we have, the better. And if we have too little information, we run the risk that the network will extract a spurious generalization. But success in learning is not merely a question of quantity of information. Quality counts too, and this is something which we will explore in several chapters. The structure of our training environment will have a great deal to do with outcome. And in some cases, it is even better to start with less, rather than more data.

A word about strategy is also appropriate here. Some tasks are better (more convincing, more effective, more informative) than others for demonstrating a point. It is easy to fall into the pitfall of "giving away the secret." This has to do with the role which is played by the output we are training the network to give us.

When we train the network to produce a behavior, we have a target output which the network learns to generate in response to a given input. This target output is called the "teacher," because it contains the information which the network is trying to learn. An important question we should ask ourselves, when we think about teaching networks a task, is whether the information represented in the teacher is plausibly available to human learners. For example, it might be interesting to teach a network to judge sentences as grammatical or ungrammatical. But we must be cautious in interpreting the results, because it is questionable that the kind of information we have provided our network (the information about when to produce an output of "ungrammatical" and when to say "grammatical") is available to children. If we believe that this information is not available, then the lessons from the network must be interpreted with caution.

Or consider the case where we are interested in the process by which children learn to segment sounds into words. Let us say we are interested in modeling this in a network in order to gain some insight into the factors which make it possible, and the way in which the task might be solved. There are several ways to model this process. One task might be to expose a network to sequences of sounds (presenting them one at a time, in order, with no breaks between words), and training the network to produce a "yes" whenever a sequence makes a word. At other times, the network should say "not yet." Alternatively, we might train a network on a different task. The network could be given the same sequences of sounds as input, but its task would be simply to predict the next sound. (This example in fact comes from one of the later exercises in the book.)

It turns out that networks can learn both sorts of tasks without too much difficulty. In the second task, however, there is an interesting by-product. If we look at the network's mistakes as it tries to predict the next sound, we find that at the beginning of a word, the network makes many mistakes. As it hears more and more of the word, the prediction error declines rapidly. But at the end of the word, when the network tries to predict the beginning of the next sound, the prediction error increases abruptly. This is sensible, and tells us that the network has learned the fact that certain sequences (namely, sequences which make up words) cooccur reliably, and once it hears enough of a sequence it can predict the rest. The prediction error also ends up being a kind of graph of word boundaries.

In both tasks the network learns about words. In the first case, it does so explicitly, from information about where words start. In the second task, the network learns about words also, but implicitly. Segmentation emerges as an indirect consequence of the task. The disadvantage of the first task is that it really does give away the secret. That is, if our goal is to understand *how* it is that segmentation of discrete words might be learned, assuming a continuous input in which segment boundaries are not marked, then we learn very little by teaching the task directly. The question is of interest precisely because the information is presumably *not* available to the child in a way which permits direct access to boundary information. Giving the information to the network thus defeats the purpose of the simulation.

Representing stimuli

We have said that a network is trained on inputs and outputs, but we have not been very specific about what those inputs and outputs look like. Network representations take the form of numbers; these may be binary (0's or 1's), integer valued (-4, 2, 34, etc.), or continuously valued (0.108, 0.489, etc.). Very frequently, information is represented by a collection of numbers considered together; these are called *vectors* and might be shown as [1 0 1 1 0]. In this case, we must have 5 input nodes; this vector tells us that the first, third, and fourth nodes are "on" (i.e., are set to 1) and the second and fifth are "off" (set to 0).

So the question is how we represent information of the sort we are interested in—words, actions, images, etc.—using numbers. This is in fact a very similar sort of problem which the nervous system solves for us. It converts sensory inputs into internal codes (neuronal firing rate, for example) which can be thought of as numerical. We might think of our problem as that of building artificial sensors and effectors for our networks.

The problem for us is that there are usually many different ways of representing inputs, and the way we choose to represent information interacts closely with the goals of the simulation. The basic issue is how much information we want to make explicitly available to the network.

At one extreme, we might wish to represent information quasi-veridically. We might present an image as a two-dimensional array of dots, in which black dots are represented as 1s and white dots are represented as 0s (or vice versa; interestingly, it doesn't matter as long as we are consistent in our mapping). This is like the half-tone format used in newspapers. If our inputs are speech, we might present the digitized voltages from an analog-digital converter, in which each input is a number corresponding to the intensity of the sound pressure wave.

Or we might choose more abstract representations. We might encode a spoken word in terms of phonetic features such as *voiced*, or *consonantal*. Each sound would be represented as a bundle of such features (a bundle being a vector), and the word would be a sequence of feature vectors.

There are two bottom lines which guide us. First, the network can only learn if it is given sufficient information. If our encoding includes information which is irrelevant, the network may learn to ignore it and solve the task using the useful information; but if the input representation does not include information which is crucial for the task, the network cannot learn at all. Second, just as was true in the design of the task, we must be concerned not to give the answer away. One goal of a simulation might be to see if a network can learn certain internal representations (say, at the level of hidden unit activation patterns). In that case, it makes little sense to go ahead and give the network the target representations as part of its input.

Choosing the right architecture

The *architecture* of a network is defined primarily by the number and arrangement of nodes in a network: How many nodes there are, and how they are interconnected. Although in theory, any task can be solved by some neural network, not any neural network can solve any task. To a large degree, the form of the network determines the class of problems which can be solved.

In this handbook we utilize two major classes of architectures: feedforward networks and simple recurrent networks. In feedforward networks, the information flows from inputs to outputs; in recurrent networks, some nodes may also receive feedback from nodes further downstream. In both cases of networks, some nodes are designated as inputs and others are designated as outputs. How many of each depends on the task and the way in which inputs and outputs are represented. (For example, if a network will be taking images as inputs and these are represented as 100x100 dot arrays, then there must be 10,000 input nodes.) Additional hidden units may also be designated.

Determining the best architecture for a task is not easy and there are no automatic procedures for choosing architectures. To a large degree, choice of architecture reflects the modeler's theory about what information processing is required for that task. As you work through the exercises, try to be aware of the differences in architecture. Ask yourself what the consequences might be if different architectures were chosen; you can even try simulations to test your predictions.

Analysis

Simulations generally involve two phases. In the first, we train a network on a task. In the second, we evaluate its performance and try to understand the basis for its performance. It is important that we anticipate the kinds of tests we plan to perform *before* the training phase. Otherwise we may find, once training is over, that we have not structured the experiment in a way which lets us ask important questions. Let us consider some of the ways in which network performance can be evaluated.

Global error. During training, the simulator calculates the discrepancy between the actual network output activations, and the target activations it is being taught to produce. Most simulators will report this error on-line, frequently summing or averaging it over a number of patterns. As learning progresses, the error will decline. If learning is perfect on all the training patterns, the error should go to zero.

But there are many cases where error cannot go to zero, even in the presence of learning. If a network is being trained on a task in which the same input may produce different outputs (that is, a task which is probabilistic or non-deterministic), then the best the network can do is learn the correct probabilities. However, there will still be some error.

Individual pattern error: Global error may also be misleading because if there are a large number of patterns (i.e., input/output pairs) to be learned, the global error (averaged over all of them) may be quite low even though some patterns are not learned correctly. If these are in fact the interesting patterns, then we want to know this. So it is important to look at performance in detail and be precise about whether the entire training set has been learned.

It may also be valuable to construct special test stimuli which have not been presented to the network during training. These stimuli are designed to ask specific questions. Does the network's performance generalize to novel cases? What in fact has the network learned? Since there are often multiple generalizations which can be extracted from a finite data set, probing with carefully constructed test stimuli can be a good way to find out what regularities have been extracted.

Analyzing weights and internal representations: Ultimately, the only way to really know what a network is doing, as well as understand how it is doing it, is to crack it open and look at its innards. This is an important advantage which network models have over human subjects; but like humans, the insides are not always easy to understand. Finding new methods to analyze networks is an area which is of great interest at the moment.

One technique which has been used profitably is hierarchical clustering of hidden unit activations. This technique can be useful in understanding the network's internal representation of patterns. After training, test patterns are presented to the network. These patterns produce activations on the hidden units, which are then recorded and tagged. These hidden unit patterns are vectors in a multi-dimensional space (each hidden unit is a dimension), and what clustering does is to reveal the similarity structure of that space. Inputs which are treated as similar by the network will usually produce internal representations which are similar—i.e., closer in Euclidean distance—to each other. Clustering measures the inter-pattern distance and represents it in tree format, with closer patterns joined lower on the tree. The partitioning of its internal state space is often used by networks to encode categories. The hierarchical nature of the category structure is captured by the subspace organization.

The limitation of hierarchical clustering is that it does not let us look at the space directly. What we might like is to be able to actually visualize the hidden unit activation patterns as vectors in hidden unit space in order to see the relationships. The problem is that it's not easy to visualize high-dimensional spaces. Methods like principal component analysis and projection pursuit can be used to identify interesting lower-dimensional "slices" in which interesting things happen. We can then move our viewing perspective around in this space, looking at activity in various different planes.

A third useful technique involves looking at activation in conjunction with the actual weights. When we look at activation patterns in a network, we are only looking at part of what a network "knows." The network manipulates and transforms information by means of the connections between nodes. Looking at the weights on these connections is what tells us how the transformations are being carried out. One of the most popular methods for representing this information is by means of what are called "Hinton diagrams" (because they were first introduced by Geoff Hinton; Hinton, 1986), in which weights are shown as colored squares with the color and size of the square indicating the magnitude and sign of the connection. Techniques involving numerical analysis have also been proposed, including skeletonization (Mozer & Smolensky 1989) and contribution analysis (Sanger, 1989).

What do we learn from a simulation?

Much of the above may still seem abstract and unclear. That's ok. Our goal here is simply to raise some questions now so that when you do the simulation exercises in the remaining chapters, you will be conscious of the design issues we have discussed here. Ask yourselves (and be critical) whether or not the simulations are framed in a way which clearly addresses some issue, whether or not the task and the stimuli are appropriate for the points that are being made, and whether you feel at the end that you have learned something from the simulation.

That's the bottom line: What have we learned from a simulation? In general there are two kinds of lessons we might learn from a simulation. One is when we have a hypothesis about how some task is carried out, and our simulation shows us that our hypothesis can in fact provide a reasonable account of the data. The second thing we may learn is new ways in which behaviors may be generated. Having trained an artificial system to reproduce some behavior of interest, our analysis of the model's solution may generate new ideas about how and why behaviors occur in humans. Of course, the model's solution need not be the human's; verifying that it is requires additional empirical work. But if nothing else, the opportunity to discover new solutions to old problems is both valuable and exhilarating!

CHAPTER 3

Learning to use the simulator

Defining the task

In this chapter, you will investigate how a neural network can be trained to map the Boolean functions AND, OR and EXCLUSIVE OR (XOR). Boolean functions just take some set of inputs, usually 1s and 0s, and decide whether a given input falls into a positive or a negative category. We can think of these Boolean functions as equivalent to the input and output activation values of the nodes in a network with 2 input units and 1 output unit. Table 3.1 summarizes the mapping con-

TABLE 3.1   The Boolean functions AND, OR and XOR

Input Activations		Output Activation		
Node 0	Node 1	Node 3		
		AND	OR	XOR
0	0	0	0	0
0	1	0	1	1
1	0	0	1	1
1	1	1	1	0

tingencies for each of these 3 functions. The first two columns of Table 3.1 specify the input activations. There are 4 possible input patterns made up from the 2^2 possible binary combinations of 0 and 1. Columns 3–5 specify the single output activation for the desired

Boolean function we require our network to compute. These input patterns and output activations define the training environment for the network we will build.

Exercise 3.1

• Do you notice anything peculiar about any of these functions? Which one would you say was the odd one out?

It may seem odd that we are asking you to perform your first simulations with abstract Boolean functions that would appear to have little bearing on our understanding of the development of psychological processes. However, we have several reasons for starting in this fashion. First, the networks that you will use to learn these functions are quite simple and relatively easy to construct. They provide a (hopefully) painless introduction to conducting your own simulations. Second, the type of networks that you will need to learn AND, OR and XOR differ in interesting and instructive ways. Many of the problems that you will encounter for these Boolean functions will illustrate some fundamental computational properties of networks that will have direct implications for your understanding of the application of networks to more complex problems. Indeed, we still often go back to these simple Boolean functions to help us work through issues which seem unfathomable in more complicated networks.

Defining the architecture

Before you can configure the simulator, you need to decide what kind of network architecture to use for the problem at hand. Let us start with the Boolean function AND. There are 4 input patterns as specified in Table 3.1 and 2 distinct outputs. Each input pattern specifies 2 activation values and each output a single activation. For every input pattern, there exists a well-defined output. So for this problem, you should use a simple feedforward network with 2 input units and a single output unit. An example of the type of network architecture you require is shown in Figure 3.1. These networks are called "single-lay-

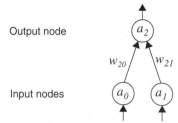

FIGURE 3.1   A single-layered perceptron for learning the Boolean function AND.

ered perceptrons" because they have single layer of weights between input and output nodes. In this book, these networks are trained with the "delta rule" described in Equation 1.5 on page 13.

We have now completed the conceptual analysis that is required to set up the simulator. You are now ready to move on to learning about some of the technical issues involved.

Setting up the simulator

The **tlearn** neural network simulator has been programmed to run on a variety of computing platforms including Macintoshes, Windows and most Unix machines that run X-windows. Furthermore, we have designed the user interface to look more or less identical across these different platforms. So the details involved in setting up your simula-

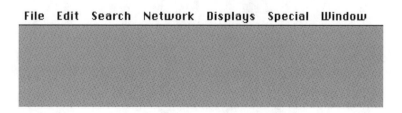

FIGURE 3.2   Startup menu for **tlearn**.

tions will be more or less the same irrespective of the type of computer system you are using. Where differences do occur, and we promise that they will be only minor differences, we will point them out. The description we give here is for the Macintosh version of the application. For information on how to install **tlearn** on your computer system, please refer to the instructions in Appendix A.

tlearn

You start up **tlearn** on your computer in the same way as any other application, such as double-clicking on the **tlearn** icon or through a sequence of keyboard commands. We will assume that you have done this. When **tlearn** has started up, your computer monitor should display a set of menus like those depicted in Figure 3.2. These menus are accessed in the standard fashion for your computer system. For example, you may click on one of the menu items and hold down the mouse button to display the options associated with that menu.

You begin the process of constructing a new network from the **Network** menu. Select the **Network** menu and drag the mouse so that the **New Project** option is highlighted as shown in Figure 3.3.

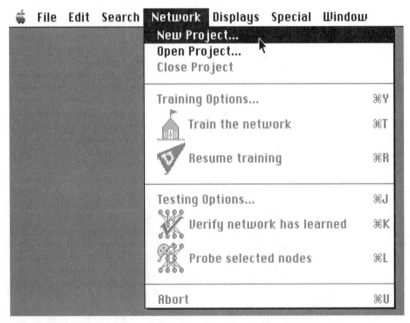

FIGURE 3.3   Select the **New Project** option from the **Network** menu.

When you release the mouse button the **New Project** dialogue box is displayed as shown in Figure 3.4. In this dialogue box you can

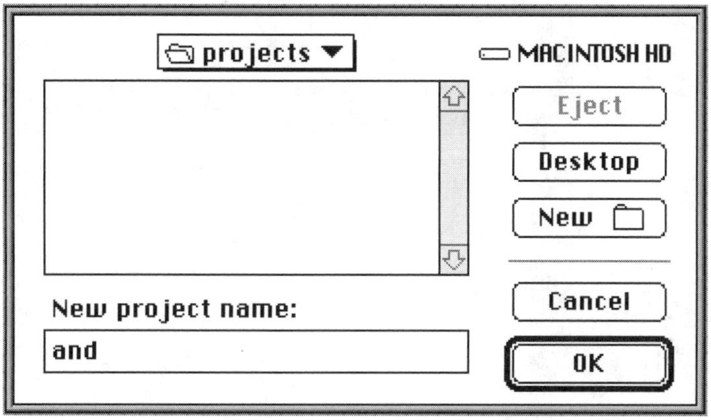

FIGURE 3.4   The **New Project** dialogue box. Select a folder or directory in which to save your current project and name the project. In this case, name the project **and.**

select a directory or folder in which to save your project files and you can name your project. In this case, call the project **and** since you are building a network to learn Boolean AND. Then click on `OK` to initiate the relevant files in your selected directory or folder. The display on your monitor should now resemble that depicted in Figure 3.5. Notice that **tlearn** has created 3 different windows—**and.cf**, **and.data** and **and.teach**. Each window will be used for entering information relevant to a different aspect of the network architecture or training environment:

- The **and.cf** file is used to define the number of nodes in a network and the initial pattern of connectivity between the nodes before training begins.

- The **and.data** file defines the input patterns to the network, how many there are and the format by which they are represented in the file.

- The **and.teach** file defines the output patterns to the network, how many there are and the format by which they are represented in the file.

By convention, **tlearn** requires that any simulation project possesses these 3 files and expects them to possess the file extensions **.cf**, **.data** and **.teach**. You can choose any name you like for the

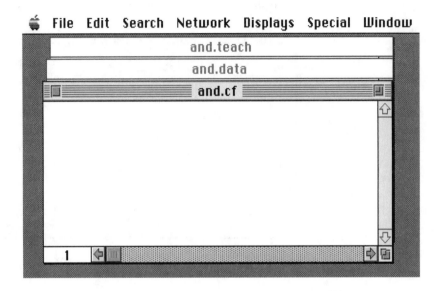

FIGURE 3.5   Startup files for a new **tlearn** project include .**cf**, .**data** and .**teach** files. The .**data** and .**teach** files define the network's training environment. The .**cf** file defines the network architecture.

and

filename. For the current project, we have chosen the filename **and**. However, all the files that belong to a project should have the same filename. Project information is stored in a special file so that you can activate previously created projects simply by activating (e.g., double-clicking) that file. The filename identifies the project file. In this case, **tlearn** has created a project file called **and** as well as the 3 required files.

Now let us consider in detail what information must be stored in the different files. It is your responsibility to enter this information. For now, we will assume that you will use the simple file creation and editing facilities that **tlearn** provides. However, you can create these files in the text editor or word processor of your choice, just so long as you remember to save the files in ASCII format.[1] The information required in the **and.cf** file is displayed in Figure 3.6. You should enter this information exactly as displayed, honoring upper case and lower case distinctions, spaces and colons. If you make a mistake while entering the characters, you can use the arrow keys on

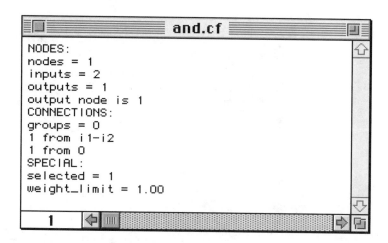

FIGURE 3.6   The **and.cf** file contains 3 sections: The NODES: section specifies the total number of units in the network and identifies which nodes play the role of input and output. The CONNECTIONS: section specifies how the units are connected to each other. The SPECIAL: section provides information that determines the initial value of the connection strengths and specifies the units whose activation values are available for inspection.

your keyboard or the mouse to navigate in the file. Use the "delete" or "backspace" key to edit out any unwanted material. When you have finished creating the file and you are sure there are no errors then save the file to disk using the standard technique on your computer system. Alternatively, you can use the **Save** command in the **File** menu of **tlearn** as shown in Figure 3.7.

1. Most word processors introduce formatting information that involve control characters like ^S which **tlearn** doesn't understand. It is imperative that you save files created by a word processor in ASCII format (also often referred to as "Text" format in many word processors) otherwise **tlearn** may fail to run. Also notice that if you create your **.cf**, **.data** and **.teach** files in another text editor, you will still have to define the project in the **New Project** dialogue box in the **Network** menu. Remember to use the same name for the project as you use for the filename in your **.cf**, **.data** and **.teach** files so that **tlearn** knows which files to look for.

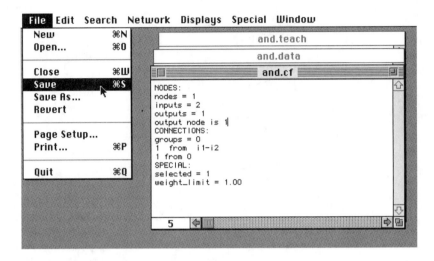

FIGURE 3.7   Saving the **and.cf** file from the **tlearn File** menu.

The Configuration (.cf) file

The **.cf** file is the key to setting up the simulator. This file describes the configuration of the network. It must conform to a fairly rigid format, but in return offers considerable flexibility in architecture. We shall now dissect the innards of the **.cf** file in some detail.

There are three sections to this file (see Figure 3.6). Each section begins with a keyword in upper case, flush-left. The three section keywords are **NODES:**, **CONNECTIONS:** and **SPECIAL:**. Note the colon. Sections must be described in the above order.

NODES: This is the first line in the file. The second line specifies the total number of nodes in the network as "**nodes = #**". *Inputs do not count as nodes.* The total number of inputs is specified in the third line as "**inputs = #**". The fourth line specifies the number of nodes which are designated as outputs according to "**outputs = #**". Lastly, the output nodes are listed specifically by number (counting the first node in the network as 1). The form of the specification is "**output nodes are <node-list>**". (If only a single output is present one can write "**output node is #**" as we have done in the **and.cf** file). *Spaces are critical.*

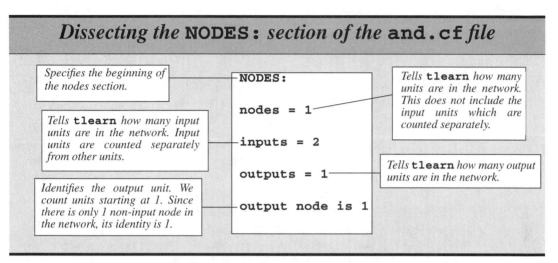

Dissecting the **NODES:** *section of the* `and.cf` *file*

Specifies the beginning of the nodes section.

Tells **tlearn** how many input units are in the network. Input units are counted separately from other units.

Identifies the output unit. We count units starting at 1. Since there is only 1 non-input node in the network, its identity is 1.

```
NODES:

nodes = 1

inputs = 2

outputs = 1

output node is 1
```

Tells **tlearn** how many units are in the network. This does not include the input units which are counted separately.

Tells **tlearn** how many output units are in the network.

CONNECTIONS: This is the first line of the next section. The line fol-

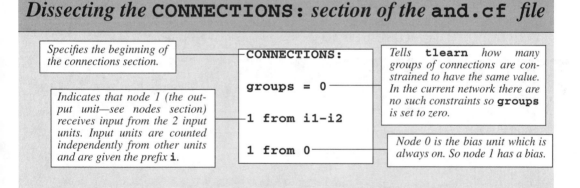

Dissecting the **CONNECTIONS:** *section of the* `and.cf` *file*

Specifies the beginning of the connections section.

Indicates that node 1 (the output unit—see nodes section) receives input from the 2 input units. Input units are counted independently from other units and are given the prefix **i**.

```
CONNECTIONS:

groups = 0

1 from i1-i2

1 from 0
```

Tells **tlearn** how many groups of connections are constrained to have the same value. In the current network there are no such constraints so **groups** is set to zero.

Node 0 is the bias unit which is always on. So node 1 has a bias.

lowing this must specify the number of groups, as in "**groups = #**". All connections in a group are constrained to be of identical strength—though in many cases "**groups = 0**" as in the **and.cf** file. Following this, information about connections is given in the form:

```
<node-list> from <node-list>
```

A **<node-list>** is a comma-separated list of node numbers, with dashes indicating that intermediate node numbers are included. A **<node-list>** contains *no spaces*. Nodes are numbered counting from 1. Inputs are likewise numbered counting from 1, but are designated as **i1**, **i2**, etc. **Node 0** always outputs a 1 and serves as the bias node. If biases are desired, connections *must* be specified from **node 0** to specific other nodes (not all nodes need be biased).

SPECIAL: This is the first line of the third and final section. Optional lines can be used to specify whether some nodes are to be linear ("**linear = <node-list>**"), which nodes are to be bipolar (meaning their values range from -1 to +1 rather than 0 to +1 which is the default "**bipolar = <node-list>**"), which nodes are selected for special printout ("**selected = <node-list>**"), and the initial weight limit on the random initialization of weights ("**weight_limit = #**"). Again, *spaces are critical*.

This may seem an overly complicated procedure for defining a network with just 2 input units and 1 output unit. You are right, it is! However, you will see that this format enables you to define far more complicated network architectures as well. So it is worth the effort to invest time to make sure you understand how the **.cf** file works.

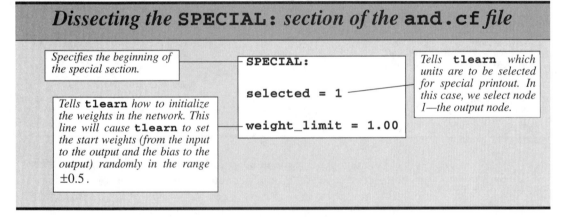

Dissecting the **SPECIAL:** *section of the* **and.cf** *file*

Specifies the beginning of the special section.

Tells **tlearn** how to initialize the weights in the network. This line will cause **tlearn** to set the start weights (from the input to the output and the bias to the output) randomly in the range ±0.5 .

```
SPECIAL:

selected = 1

weight_limit = 1.00
```

Tells **tlearn** which units are to be selected for special printout. In this case, we select node 1—the output node.

The Data (.data) file

The **and.data** file defines the input patterns which are presented to **tlearn**. Enter the data as shown in Figure 3.8. Don't forget to save the file when you have finished.

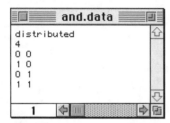

FIGURE 3.8   The **and.data** file.

The first line must either be **distributed** (the normal case) or **localist** (when only a few numbers of many input lines are non-zero). The next line is an integer specifying the number of input vectors to follow. The remainder of the **.data** file consists of the input. These may be input as integers or floating-point numbers.

In the (normal) **distributed** case, the input is a set of vectors. Each vector contains *i* floating point numbers, where *i* is the number of inputs to the network. Spaces between the numbers are critical. In the **localist** case, the input is a set of **<node-list>**'s listing only the numbers of those nodes whose values are to be set to one. Node lists follow the conventions described in the **.cf** file.

The Teach (.teach) file

The **.teach** file is required whenever learning is to be performed. Enter the data for the **and.teach** file as shown in Figure 3.9. Don't forget to save the file when you are finished. As with the **.data** file, the first line must be either **distributed** (the normal case) or **localist** (when only a few of many target values are nonzero). The next line is an integer specifying the number of output vectors to follow. The ordering of the output patterns matches the ordering of the corresponding input patterns in the .**data** file. In the (normal) **distributed** case, each output vector contains *o* floating point or integer numbers, where *o* is the number of outputs in the network. An

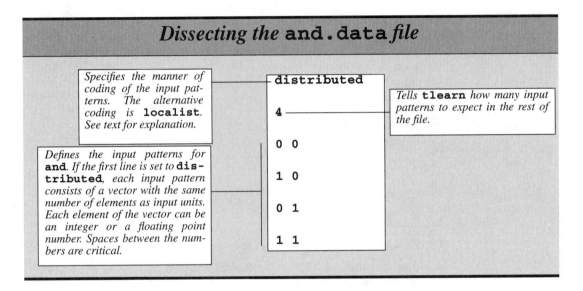

Dissecting the **and.data** file

Specifies the manner of coding of the input patterns. The alternative coding is **localist**. See text for explanation.

Defines the input patterns for **and**. If the first line is set to **distributed**, each input pattern consists of a vector with the same number of elements as input units. Each element of the vector can be an integer or a floating point number. Spaces between the numbers are critical.

```
distributed
4
0 0
1 0
0 1
1 1
```

Tells **tlearn** how many input patterns to expect in the rest of the file.

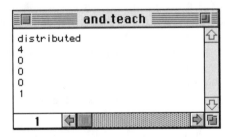

FIGURE 3.9   The **and.teach** file.

asterisk (*****) may be used in place of a floating point number to indicate a "don't care" output. In the **localist** case, the input is a set of **<node-list>**'s listing only the numbers of those nodes whose values are to be set to one. Node lists follow the conventions described in the **.cf** file.

Checking the architecture

If you have typed in the information to the **and.cf**, **and.data** and **and.teach** files correctly then you should experience no problems running the simulation. However, **tlearn** offers a useful way to check whether the **and.cf** file has been correctly specified. You can

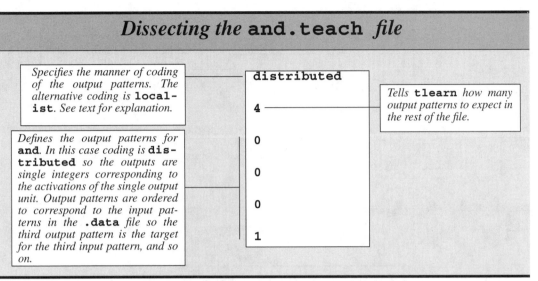

Dissecting the `and.teach` *file*

Specifies the manner of coding of the output patterns. The alternative coding is **localist**. See text for explanation.

Tells **tlearn** how many output patterns to expect in the rest of the file.

Defines the output patterns for **and**. In this case coding is **distributed** so the outputs are single integers corresponding to the activations of the single output unit. Output patterns are ordered to correspond to the input patterns in the **.data** file so the third output pattern is the target for the third input pattern, and so on.

```
distributed

4

0

0

0

1
```

display a picture of your network architecture using the **Network Architecture** option in the **Displays** menu. The architecture for Boolean AND is shown in Figure 3.10. The buttons at the top of the display enable you to adjust your view of the network. For example, it is possible to view the network without the bias displayed simply by clicking on the **Bias** button. Note, however, that the adjustments you make to the display do not effect the contents of the network configuration file. **tlearn** will not complain if there are any mistakes in the training files (**.data** and **.teach**) when you display the network architecture. An error message will be displayed when you attempt to train the network, if there is a mistake in the syntax of your training files. Needless to say, **tlearn** cannot identify incorrect entries in the training set data. If you accidently enter the input pattern 0 1 instead of 1 1, you will receive no notification. Check your network files carefully!

Running the simulation

Once you have specified the three input files and saved them to disc, you are almost ready to run your first simulation. First, however, you must specify a number of parameters for **tlearn** that will determine the initial start state of the network as well as the learning rate and

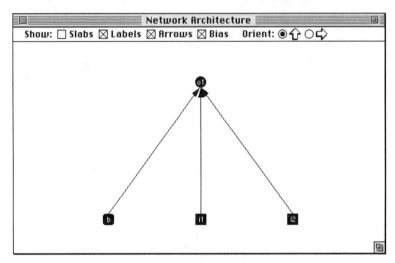

FIGURE 3.10   The **Network Architecture** display. **tlearn** reads the information in the **and.cf** file to construct this display. The user can customize the network display using the buttons at the top of the display. These changes do not effect the contents of the network configuration file.

momentum. For a brief overview of the kind of parameters that you can set in **tlearn** consult Figure 3.11. To activate this dialogue box you need to select the **Training options...** option in the **Network** menu. First, we indicate the number of training sweeps that the network should perform before halting. A training sweep consists of a single presentation of an input pattern causing activation to propagate through the network and the appropriate weight adjustments to be carried out. In Figure 3.11 the number of training sweeps has been set to 1000 which means that 1000 patterns are to be presented to the network. The order in which patterns are presented to the network is determined by the **Train sequentially** and **Train randomly** buttons. Activate the **Train sequentially** button (by clicking it with the mouse) to present patterns in the order in which they appear in the **.data** and **.teach** files. Activate the **Train randomly** button to present patterns in random order. In Figure 3.11 **tlearn** has been set to select patterns at random.

The initial state of the network is determined by the weight values assigned to the connections before training begins. Recall that the **.cf** file specifies a **weight_limit** in the **SPECIAL:** section (see

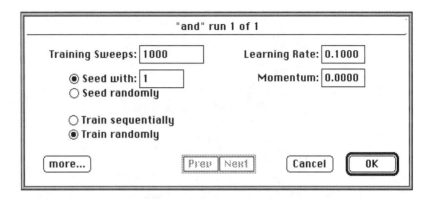

FIGURE 3.11   The **Training Options** dialogue box activated from the **Net-work** menu. Here you specify many of the parameters for train-ing the network including the number of sweeps, learning rate, momentum, the initial startup configuration of the weights and the manner in which patterns are selected for training.

Figure 3.6). Weights can be assigned according to a random seed indi-cated by the number next to the **Seed with:** button. You can select any number you like. The advantage of this approach is that a simula-tion can be replicated using the same random seed—meaning that the initial start weights of the network will be identical and patterns will be sampled in the same random order. Make sure that you activate the **Seed with:** button if you wish to use this method. Alternatively, you may wish to let the computer choose a random seed for you in which case activate the **Seed randomly** button. Note that *both* of these random seed procedures select a set of random start weights within the limits specified by the `weight_limit` parameter in the `.cf` file. The only difference between the procedures is that in one case you have control over the random seed and in the other case you relinquish control to the computer. In Figure 3.11 we have specified a random seed of 1. Make sure you choose the same random seed for purposes of this demonstration.

The other parameters specified in Figure 3.11 include **Learning Rate:** and **Momentum:** You have already met the learning rate parameter in Chapter 1. It determines how fast the weights are changed in response to a given error signal. The momentum parame-ter has not yet been properly introduced. For the moment, set these parameters to the values specified in Figure 3.11, i.e.,0.1 and 0.0.

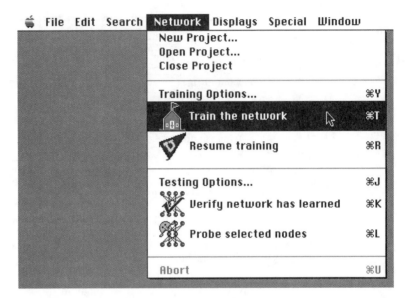

FIGURE 3.12   Use **Train the network** to begin a simulation. Alternatively, use the keyboard shortcut ⌘-**T**

There are many other parameters that can be set through the **Training options** dialogue box. For the time being, however, just click ⬚ OK ⬚ to accept the current configuration and close the dialogue box.

Now you can start to train the network. You can do this from the **Network** menu by choosing the option **Train the network** as shown in Figure 3.13. Immediately, you start training the network the **tlearn Status** display appears—see Figure 3.13 (a). The status display indicates how many sweeps have been completed and provides the opportunity to abort training, dump the current state of the network in a weights file and iconify the status display to clear the screen for other tasks while **tlearn** runs in the background. The status display also indicates when **tlearn** has completed the current round of training—see Figure 3.13 (b).

Now that you've trained the network for 1000 sweeps, let's see what it's learned. There are a variety of ways to determine whether the network has solved the problem you have set it. We will examine two methods here:

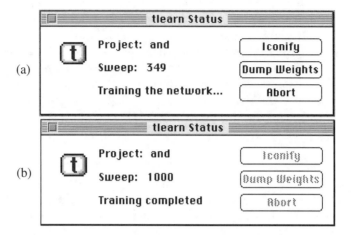

(a)

(b)

FIGURE 3.13   The **tlearn Status** display window indicates when the simulator is occupied training the network and when it has completed its training cycle. During training it is possible to Iconify the status window and revert to another task (while training continues in the background). Training can be terminated by clicking on the **Abort** button. Weights files can be saved by **Dump**ing **Weights**.

1.   Examine the global error produced at the output nodes averaged across patterns.

2.   Examine the response of the network to individual input patterns.

Global error

For each input pattern, the network will produce an output response. As we saw in Chapter 1 the error at the output can be calculated simply by subtracting the actual response from the desired or target response. The value of this discrepancy can be either positive or negative—positive if the target is greater than the actual output and negative if the actual output is greater than the target output. We wish to determine the global performance of the network across all four of the input/output pairs that make up Boolean AND.

Let us define global error as the average error across the four pairs at a given point in training. In fact, **tlearn** provides a slightly more complicated measure of global error called the RMS or Root Mean

Square error. To determine the RMS error **tlearn** takes the square of the error for each pattern, calculates the average of the squared errors for all the patterns and then returns the square root of this average. Using the RMS error instead of a raw average prevents **tlearn** from cancelling out positive and negative errors. The calculation can be expressed more succinctly in mathematical notation as:

$$\text{rms error} = \sqrt{\frac{\sum_k (\vec{t}_k - \vec{o}_k)^2}{k}} \qquad \text{(EQ 3.1)}$$

In Equation 3.1 the symbol k indicates the number of input patterns and $\vec{o}_k$ is the vector of output activations produced by input pattern k. The number of elements in the vector corresponds to the number of output nodes. Of course, in the current problem—Boolean AND—there is only one output node so the vector $\vec{o}_k$ contains only one element. The vector $\vec{t}_k$ specifies the desired or target activations for input pattern k.

Exercise 3.2

- What value should k take for Boolean AND in Equation 3.1?

---

tlearn keeps track of the RMS error throughout training. The easiest way to observe how RMS error changes is through the **Error Display**. Activate the **Error Display** from the **Displays** menu. An **Error Display** for the simulation we've just run is shown in Figure 3.14. Error is reported on the graph every 100 sweeps. The x-axis on the graph indicates the number of sweeps and the y-axis the RMS error. If you prefer error to be displayed by a continuous line then click on the **Line** button.

Notice that error decreases as training proceeds such that after 1000 sweeps RMS error ≈ 0.35. What does this error level mean? It indicates that the average output error is just 0.35. So the output is off target by approximately 0.35 averaged across the 4 patterns. It would appear that the network has not solved the problem yet. In fact, this level of error may reveal that the network has solved the AND problem. It all depends on how we define an acceptable level of error. Unfortunately, it is not always easy to evaluate network performance

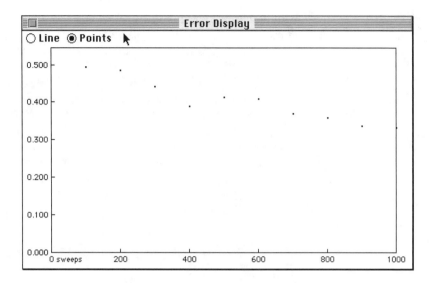

FIGURE 3.14 The error display for global error. The x-axis represents the number of sweeps and the y-axis the RMS error. The error is reported every 100 sweeps and is indicated by a dot. The dots can be joined to form a line by clicking on the **Line** button.

on the basis of global error alone. Although the network may have a low RMS error, there is no guarantee that the network has categorized all the input patterns correctly (see Exercise 3.3).

Exercise 3.3

1. How many times has the network seen each input pattern after 1000 sweeps through the training set?

2. How small must the RMS error be before we can say the network has solved the problem?

Pattern error

Recall that RMS error reflects the average error across the 4 input patterns. It is difficult to know whether the error is uniformly distributed across different patterns or whether some patterns have been learned correctly while others remain incorrectly learned. In order to distinguish between these possibilities, we need to examine the output activations for individual patterns. **tlearn** provides several facilities for viewing the activation values of individual nodes. We will consider two of them.

The most accurate method for determining pattern output is to present each input pattern to the network, just once, and observe the resultant output node activations. The output activations can then be compared with the teacher signal in the **.teach** file. Present each input pattern to the network by selecting the **Verify network has learned** from the **Network** menu. The **tlearn** status window will indicate that it has conducted 4 sweeps (one for each training input) and that **Verification** is **Completed**. At the same time a new window called **Output** will open as shown in Figure 3.15. **Output** indi-

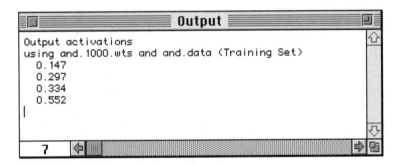

FIGURE 3.15   The **Output** activation window. The activation for each output node is displayed one pattern per row. There is only one output node so only one activation value is displayed. Activations are based on the most recent state of the network as recorded in **and.1000.wts**.

cates that it has used the file **and.1000.wts** as a specification of the state of the network (we will examine the **.wts** file in more detail shortly) and that it has used the **and.data** training patterns to verify network performance. **Output** then displays output activations one pattern per row. Since there is only one output node there will only be a single activation on each row. You can compare the activation values in Figure 3.15 to the target activations specified in your **and.teach** file.

Exercise 3.4

---

1. Has **tlearn** solved Boolean AND?

2. Calculate the exact value of the RMS error and compare it to the value plotted in Figure 3.14.

A shortcut method exists for observing output activations for individual input patterns. This facility is accessed through the **Node acti-vations** option in the **Displays** menu. Activating this option will yield the display in Figure 3.16. The **Node Activations Display**

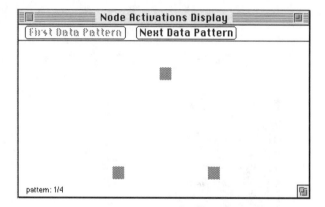

FIGURE 3.16   The node activation display. Click on the **First Data Pat-tern** button to view node activations when the first input pattern is presented. The size of the white square indicates level of activation. Grey squares indicate zero activation. The current pattern (input pattern 1 of 4) has 0 0 as its input and 0 (almost) as its output. This is a correct response. Scroll through the other patterns by clicking on **Next Data Pattern**.

shows the activations of the two input nodes and the output node. Activation levels are indicated by white bordered squares. Large white squares indicate high activations. Small white squares indicate low activations. A grey square indicates an inactive node. In Figure 3.16 the two input nodes are inactive—corresponding to the input pattern 0 0—and the output node is only slightly active, i.e., more off than on. You can display node activations for the other three input patterns simply by clicking on the **Next Data Pattern** button. **tlearn** will then display activation values in the order they are listed in the training files.

Examining the weights

Input activations are transmitted to other nodes along modifiable connections. The performance of the network is determined by the strength of the connections—also called their weight values. To understand how the network accomplishes its task, it is important to examine the weights in the network. **tlearn** offers several options for displaying the weights in the network. The shortcut method utilizes the **Connection Weights** option in the **Display** menu. The connections weights display (shown in Figure 3.17) depicts the weight

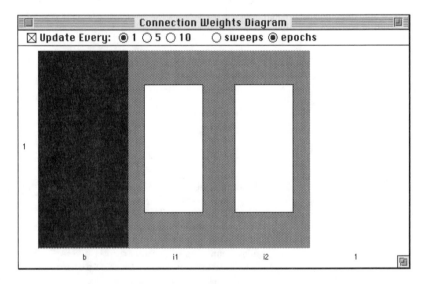

FIGURE 3.17   The **Connection Weights** display uses a Hinton diagram to plot the *relative* strength of each of the weights (including bias weights) in the network. Black rectangles signify negative weights and white rectangles signify positive weights. Weights are identified by their row and column in the Hinton diagram matrix (see text for further explanation).

values as white (positive) or black (negative) rectangles. The size of the rectangle reflects the absolute size of the connection. In the literature, these displays are usually called "Hinton diagrams." The array, or matrix, of rectangles are organized in rows and columns. You read Hinton diagrams in accordance with this row/column arrangement. All the rectangles in the first column code the values of the connections emanating from the bias node. The rectangles in the second column code the connections emanating from the first input unit. As we read across the columns we observe the connections emanating from higher numbered nodes (as defined by the `.cf` file). The rows in each column identify the destination node of the connection. Again, higher numbered rows indicate higher numbered destination nodes. In the current example, there is only one node that receives inputs (the output node). All the other nodes are input nodes themselves and so by definition receive no incoming connections. The large black rectangle in the first column refers to the value of the weight connecting the bias to the output node. The smaller white rectangle in the second column codes the connection from the first input node to the output node. The slightly larger white rectangle in the third column codes the connection from the second input node to the output node.

Exercise 3.5

•   Why do you think the fourth column is empty?

The Hinton diagram in Figure 3.17 gives us a fairly good idea as to how the network has solved Boolean AND. Notice that the bias has a strong negative connection to the output node while the two input nodes have moderately sized positive connections to the output node. This means that one active input node by itself cannot provide enough activation to overcome the strong negative bias and turn the output node on. However, two active input nodes together can overcome the negative bias. This situation is exactly what we need to solve Boolean AND: The output node only turns on if both input nodes are active.

Sometimes, we need a more accurate picture of the internal structure of the network. For example, we might need to know the exact value of the weights in the network. **tlearn** keeps an up-to-date record of the network's state in a weights file. These files are saved on the disk at regular intervals (which you, the user, can specify). **tlearn** also saves a weights file at the end of each training session. You will find the current weights file in the same folder as your project files. The weights file should be called **and.1000.wts**. You can

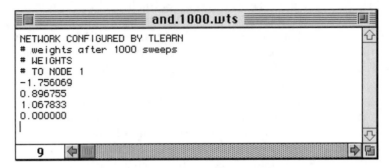

FIGURE 3.18   The weight configuration stored after 1000 sweeps of training in the file **and.1000.wts**. The file lists the connections to each receiving node. For each receiving node, input connections are listed starting with the bias, then the input nodes and all other nodes in the network in ascending order as identified in the **.cf** file.

Open this file from the **File** menu in **tlearn**. The contents of the file are shown in Figure 3.18. The file lists all the connections in the network grouped according to receiving node. In the **and.cf** file only one receiving node is specified—the output node or node 1. Connections going into each node are listed from the bias node through the input nodes to the higher numbered nodes as specified in the **.cf** file. The first number in the **and.1000.wts** file (after the **# TO NODE 1** line) represents the weight on the connections from the bias node to the output node. The second number (0.896755) shows the connection from the first input node to the output node. The third number shows the connection from the second input node to the output node. The final number (0.000000) shows the connection from the output node to itself. Recall that this connection is non-existent—we are using a simple feedforward network here. Nevertheless, the format of the weights file is defined such that all possible input connections from every potential sending node are specified. With more complicated network architectures you will find that this seemingly unnecessary complexity has some saving graces!

Network training can be continued by selecting the **Resume training** option on the **Network** menu. **tlearn** will automatically extend training by another 1000 sweeps and adjust the error display to accommodate the extra training sweeps. Try training the network for

Exercise 3.6

- Now that you know the precise configuration of the network, calculate by hand output activations for each input pattern and see if you can confirm the network's calculations as depicted in the **Output** window (see Figure 3.15).

---

an extra couple of thousand sweeps and observe whether the RMS error decreases significantly. Then practice the techniques you have learned in this section for evaluating performance and the state of the network.

The role of the start state

The network solved Boolean AND starting with a particular set of random weights and biases. Now run the simulation again but with a different set of initial weights and biases. This is easy to do. Just use a different random seed. Open up the **Training Options** dialogue box and select a different random seed (say 2). When you issue the command **Train the network, tlearn** wipes out the learning that has taken place in the network and provides a new set of random weights determined by the random seed you have used. Training continues in the usual fashion. You can resume training beyond the specified number of sweeps using the **Resume training** option.

Exercise 3.7

---

1. Train the network using a variety of different random seeds. Does the network show the same pattern of error for all the random start states?

2. Does the error always decrease from one point on the error plot to the next?

Exercise 3.7

---

3. If two simulations exhibit the same level of error at the end of training, does this mean that the connections weights in the network are identical?

4. Does the size of the weight limit parameter (specified in the `.cf` file) influence the outcome of network training? Try running a simulation with a large weight limit such as 4.0 .

Start states can have a dramatic impact on the way the network attempts to solve a problem and on the final solution it discovers. Researchers often attempt to replicate their simulations using different random seeds to determine the robustness or reliability of network performance on a given type of problem. Training networks with different random seeds is like running subjects on experiments. You repeat the experiment to determine the degree to which the outcome depends upon the participating individual or other factors of interest. Alternatively, running a simulation with different random seeds might be likened to evaluating the fitness of different organisms to adapt to an environmental niche. The initial state of the network corresponds to the organism's phenotype.

The role of learning rate

Recall that you specified the **Learning rate** parameter in the **Training options** dialogue box to be 0.1. Now investigate the impact of the learning rate for the network's performance on Boolean AND. Learning rate determines the proportion of the error signal (or more accurately δ) which is used to change the weights in the network. Large learning rates lead to big weight changes. Small learning rates lead to small weight changes (see Equation 1.5 on page 13). In order to examine the effect of learning rate on performance, you need to run the simulation in such a fashion that learning rate is the only factor that has been changed. In particular, you need to start the network off in the same state as before, i.e., with the same set of random weights and biases. Then you can compare the results of the new simulation with your previous run. Again notice the similarity to running controlled experiments. We know that the start state of the network can have a dramatic effect on learning so we avoid confounding experimental variables by using an identical start state.

Open the **Training Options** dialogue box and select a random seed of 1. This ensures that the network starts off in an identical fashion to previous runs we have observed. Next set the **Learning rate** parameter to 0.5. Finally, make sure that you have selected the **Train Randomly** button. Close the dialogue box and **Train the network**.

Exercise 3.8

- Do you notice any difference from the previous run in the final solution that the network discovers? What is the effect of changing the learning rate? Try repeating this experiment with two or three other values of the learning rate parameter.

Generally, modelers use a small learning rate to avoid large weight changes. Large weight changes made in response to one pattern can disrupt changes made in response to other patterns so that learning is continually undone on consecutive pattern presentations. In addition, large weight changes can be counter-productive when the network is close to a solution. The weights may end up in a configuration which is further away from the optimal state than it was prior to the weight change!

Logical Or

Boolean AND is solved relatively quickly by a single-layered perceptron across a fairly wide-range of learning conditions (start state, learning rate, etc.). Now evaluate the network's capacity to master Boolean OR—the second of the Boolean functions listed in Table 3.1.

Exercise 3.9

- What type of network architecture should you use for Boolean OR?

You can set up the files necessary for running an Or simulation by opening the **New project...** option in the **Network** menu. Three files will be opened by `tlearn`—`or.cf`, `or.data` and `or.teach`. Initially, they will be empty files. Configure these files to contain the same information as the corresponding **and** files, except for the `or.teach` in which the target patterns should be those indicated in the output activations of the Or column in Table 3.1. If you wish, you can just copy all the **and** files and rename them with their appropriate **or** titles. When you open a **New Project...** entitled `or`, `tlearn` will use the duplicate files you have created. You will then only need to edit the `or.teach` file.

Exercise 3.10

---

1. Before you set up the network and run the simulation, can you predict how the network will attempt to solve this problem?

2. What does the `or.teach` file look like?

Activate the **Error Display** and then **Train the network** using the same **Training options...** that you used for the initial run on Boolean AND (see Figure 3.11). Remember to set the training options before you attempt to train the network. Use a random seed of 1, a learning rate of 0.1 and select patterns to **Train randomly**. In this manner, we can compare network performance on Boolean OR and AND directly. All we've changed is the target output activations. Let's evaluate network performance after 1000 sweeps of training. The RMS error curve is shown Figure 3.19. Compare this with the performance on Boolean AND depicted in Figure 3.14. The error decreases faster and to a lower level by 1000 sweeps. This suggests that the network might have solved Boolean OR if we evaluate error in accordance with the "rounding-off" criterion (see Exercise 3.3 on page 63). However, as we saw in the section on **Pattern Error** on page 49, the RMS error may provide a misleading picture of network performance. It is necessary to examine the errors on individual patterns to be confident that the network has indeed solved the problem.

Earlier in the chapter (see Exercise 3.4 on page 65), we used the **Verify network has learned** option in the **Network** menu to obtain a set of output activations for the network trained on Boolean

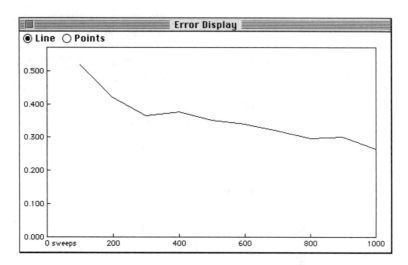

FIGURE 3.19   Network performance over 1000 sweeps of training on Boolean OR.

AND. We used a rounding-off method to compare output activations with the target activations and thereby determined whether the network had solved the problem. However, **tlearn** provides additional techniques for evaluating the error on individual patterns. Select **Testing Options...** in the **Network** menu. The dialogue box shown in Figure 3.20 will appear. **Testing Options...** sets a variety of parameters for evaluating network performance without further learning taking place when the **Verify network has learned** or **Probe selected nodes** options in the **Network** menu is selected. Take some time to browse through the options in this dialogue box and see if you can guess their functions.

We want to test the network's output for individual patterns so we will eventually use the **Verify network has learned** option in the **Network** menu. By default, this option tests the network's response to the input patterns in the training set using the current state of the weights in the network. In Figure 3.20 you can see that the **Weights file:** option is selected for the **Most recent** weights file, i.e., the current state of the network. This is just the weights file **or.1000.wts** which **tlearn** selects automatically. If you wanted you could select another weights file by selecting the **Earlier one:** button and typing the name of an appropriate weights file in the text

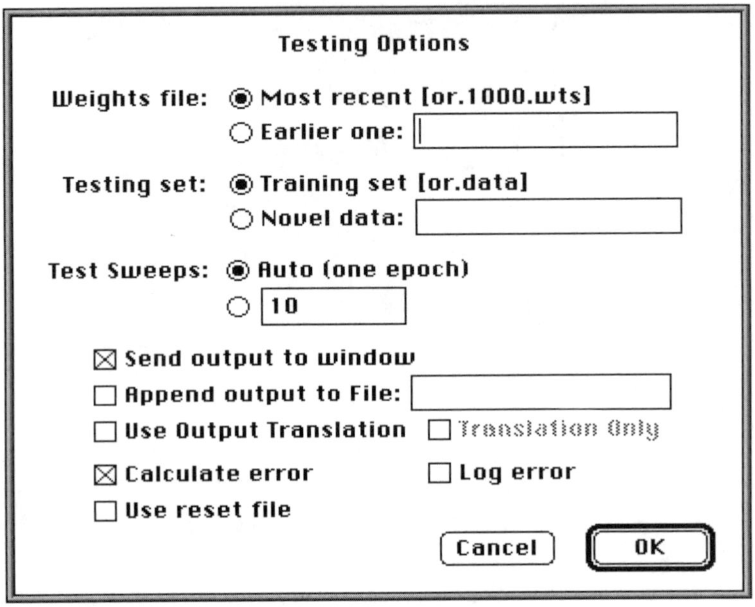

```
┌─────────────────────────────────────────────────────────┐
│                    Testing Options                        │
│                                                           │
│  Weights file:  ⦿ Most recent [or.1000.wts]              │
│                 ○ Earlier one: [│                      ]  │
│                                                           │
│  Testing set:   ⦿ Training set [or.data]                 │
│                 ○ Novel data: [                       ]  │
│                                                           │
│  Test Sweeps:   ⦿ Auto (one epoch)                       │
│                 ○ [10        ]                            │
│                                                           │
│         ☒ Send output to window                           │
│         ☐ Append output to File: [                    ]  │
│         ☐ Use Output Translation  ☐ Translation Only      │
│         ☒ Calculate error          ☐ Log error           │
│         ☐ Use reset file                                  │
│                              [ Cancel ]  [[  OK  ]]       │
└─────────────────────────────────────────────────────────┘
```

FIGURE 3.20   The **Testing Options...** network dialogue box which is accessed from the **Network** menu.

box. Alternatively, you can double-click on the text box to bring up a list of all the weights files in your current folder. This is a useful aid to memory if you can't remember the names of your weights files but make sure you select a weights file that is appropriate to your current network architecture. So far this isn't a problem because we've always used networks with identical architectures. But once you start using networks that vary in the number of connection weights, it will be crucial to select the right type of weights files. **tlearn** will not complain that you've chosen the wrong file. It will even produce output activations when you **Verify network has learned**! Load your weights files carefully.

The second decision to make in the **Testing Options...** dialogue box concerns the **Testing set:** Do you want to evaluate performance on the patterns in the training set or the response of the network to an entirely new set of patterns. The latter can be important for discovering how the network generalizes (more of that in Chapter 6). For the moment, we are interested in performance on the training set so select the **Training set** button. Make sure that `tlearn` has selected the correct training set—in this case `or.data`.

Next set **Test Sweeps:** to **Auto (one epoch)**. This tells `tlearn` to present each training pattern to the network just once and in the order specified in the `.data` file when we select the **Verify network has learned** option. Finally, set all the other square-boxes at the bottom of the dialogue box in Figure 3.20 as indicated. In particular, make sure that the **Calculate error** box is on. This tells `tlearn` to display the error for individual patterns when you choose the **Verify network has learned** option. The **Error Display** used for the RMS error plot is also used for this option.

When you have selected the relevant options, close the **Testing options...** dialogue box and select **Verify network has learned**. If all is well, you should see an **Error Display** like that in Figure 3.21. Performance on all patterns is within criterion though

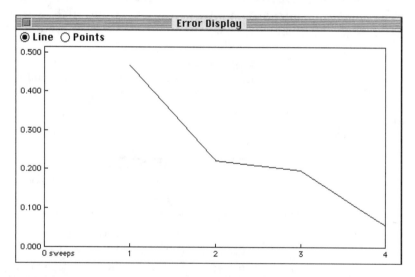

FIGURE 3.21   Error display for individual patterns in Boolean OR.

the error is least on pattern 4 (1 1). You might like to train the network further (using the **Resume training** command) to see how quickly the error reduces on the other training patterns.

Exercise 3.11

- How did the network solve Boolean OR? Use the techniques that we reviewed in Exercise 3.6 on page 66 to answer this question. Is the network sensitive to training with different random seeds?

Exclusive Or

Create a new set of files for the EXCLUSIVE OR problem shown in Table 3.1. Create the project file by selecting the **New Project** option in the **File** menu. Call the project **xor**. If you have already copied the files over from the **or** project to create **xor.cf**, **xor.data** and **xor.teach**, all you need to do is edit the **xor.teach** file so that it has the right outputs as specified in Table 3.1. Otherwise, you will need to create the network files from scratch.

Once you have created the necessary files, set the training options so that they conform to the parameters that you have used previously for **and** and **or**. Now train the network for 1000 sweeps. Evaluate performance on XOR.

Exercise 3.12

- Has the network solved XOR? If not, try **Resume training** for a further 4000 sweeps. Does the network solve the problem?

You will discover that your network architecture has considerable difficulties solving XOR over a wide range of start states (random seeds) and learning rates. In fact, a single-layered perceptron is unable to solve XOR. The difficulty arises with the non-linearity inherent in the Boolean mapping. In the next chapter we will examine the type of network XOR architecture and training regime that permits the solution of this problem.

Answers to exercises

Exercise 3.1

• The set of input patterns are identical for all the Boolean functions represented in Table 3.1. AND, OR and XOR differ in the set of responses to the 4 inputs. For example, AND demands the response 0 to the input pattern 0 1 but OR demands a response of 1 to the same input pattern. Therefore, we must expect our network to exploit a different set of connection weights depending upon which Boolean function it is attempting to compute.

The output responses required for the Boolean functions AND and OR increases with the amount of activity at the input. So the input 0 0 produces 0 output for AND and OR while the input 1 1 produces an output of 1. In contrast, for the Boolean function XOR, there is a non-linear relation between the amount of activity at the input and the activity at the output—the inputs 0 0 and 1 1 both produce the output 0. In this respect, XOR is the odd-one-out of these 3 Boolean functions. This fact will have important implications for network training as we shall see in "Exclusive Or" on page 62.

Exercise 3.2

• There are 4 different input patterns so the value of $k = 4$.

Exercise 3.3

1. There are 4 input patterns in the training environment so within 1000 sweeps the network will see each pattern 250 times. This is, in fact, only an approximation since the simulator selects patterns at random from the training set. However, for increasing numbers of training cycles, the percentage difference in selection of given patterns diminishes. In other words, the patterns in the training set are selected with almost equal frequency. For the statistically minded reader, patterns are selected from the

training set *with replacement*. You can force the network to select randomly *every* pattern in each training epoch by deactivating the **With replacement** check box in the enlarged **Training options** dialogue box. Click on the **more** button to reveal a new universe of training options!

2. The answer to this question may seem obvious: The network has solved the problem when the RMS error has been reduced to zero. However, there are several complications which mitigate against this solution. First, recall that the activation function of the output unit is the sigmoid function defined in Equation 1.2 and depicted graphically in Figure 1.3 on page 5. The net input to the node determines the node's activation. You can observe in Figure 1.3 that the activation curve never quite reaches 1.0 nor reduces to 0.0. In fact, to achieve these values the net input to the node would need to be $\pm\infty$ (infinity) respectively. Nodes never receive $\pm\infty$ input so there will always be a residual finite error. The question then arises as to what level of error is acceptable.

There is no single correct answer. For example, you might suggest that all outputs should be within 0.1 of their target. So if the target is 0 then outputs > 0.1 should be considered incorrect and if the target is 1 then outputs < 0.9 should be considered incorrect. However, these criteria have been set in an arbitrary fashion. Why not choose a criterion of 0.2 rather than 0.1? An alternative solution would be to round off the activation values. Activations closest to 1.0 are judged to be correct if the target is 1.0. Activations closest to 0.0 are judged to be correct if the target is 0.0. In effect, this alternative sets the criterion to 0.5. To repeat, there is no *right* answer in setting the error criterion. Many researchers accept a rounding off procedure. Others set more stringent demands. It can be useful therefore when evaluating the performance of the network to determine how different values of the error criterion effect the picture of performance.

Let us assume for the time being that we use rounding off to determine whether the output is correct or incorrect. We want to know what level of global RMS error *guarantees* that all the patterns have been learned to criterion. Consider the case where all patterns are 0.5 off their targets, i.e. $|t - o| \leq 0.5$. Substituting in Equation 3.1 gives an overall RMS error of 0.5:

$$\text{rms error} = \sqrt{\frac{0.5^2 + 0.5^2 + 0.5^2 + 0.5^2}{4}} = 0.5 \qquad \textbf{(EQ 3.2)}$$

However, it is also possible to observe a RMS error ≤ 0.5 in which some of the patterns are still categorized incorrectly. For example, errors could be 0.6, 0.6, 0.1 and 0.1 to yield an RMS error of 0.43. So we need a more conservative error level to guarantee that all the patterns are correct. We can be confident that all the patterns are within criterion when only one pattern is contributing substantially to the error. This means that a maximum

$$\text{RMS error} \leq \sqrt{\frac{(0.5^2)}{4}} = 0.25$$

guarantees that all the patterns have been categorized correctly. Of course, this value holds only for binary targets to 4 training inputs. Different levels of global error will be appropriate for other problems.

Exercise 3.4

1. If we use a rounding-off method to evaluate network performance then we get the output values indicated in Table 3.2. The exact output values are indicated in the **Output** column and their rounded values in the

TABLE 3.2   Activation values and errors in Boolean AND.

Input		Output	Rounded Off	Target	Squared Error
0	0	0.147	0	0	0.022
1	0	0.297	0	0	0.088
0	1	0.334	0	0	0.112
1	1	0.552	1	1	0.201
				RMS Error	0.323

Rounded Off column. The **Target** scores match the rounded values exactly. So by this error criterion, **tlearn** has solved Boolean AND.

2. The exact RMS error is calculated according to Equation 3.1 (see page 48). In Table 3.2 the **Squared Error** for each input pattern is calculated by squaring the difference between the output and target values. The RMS error is just the square root of the average of all the squared

errors, i.e., 0.323 . Notice this error score is identical to the error plotted by **tlearn** in Figure 3.14 after 1000 sweeps of training. This should come as no surprise. The two error scores should be identical!

Exercise 3.5

- The fourth column codes the connection from the fourth node in the network to the output unit. In this case, the fourth node is just the output node itself. The first three nodes were the bias and the two input nodes. So the fourth column depicts the connection between the output node and itself—a recurrent connection. The network we are currently examining is a feedforward network. There are no recurrent connections so the fourth column is empty.

Exercise 3.6

- To calculate the activation of the output node for each input pattern we need to find the sum of the weighted input activations and use the sum as the input to the logistic activation function (see Exercise 1.1 on page 8). We'll display these calculations in table format—one section of the table

TABLE 3.3   Calculating output node activations

		Input Activation	Connection Strength	Weighted Activations	Sum of Weighted Activations	Logistic of Net Input
Pattern 1	Bias	1	-1.756069	-1.756069	-1.756069	**0.147**
	Input One	0	0.896755	0		
	Input Two	0	1.067833	0		
Pattern 2	Bias	1	-1.756069	-1.756069	-0.859314	**0.297**
	Input One	1	0.896755	0.896755		
	Input Two	0	1.067833	0		
Pattern 3	Bias	1	-1.756069	-1.756069	-0.688236	**0.334**
	Input One	0	0.896755	0		
	Input Two	1	1.067833	1.067833		

TABLE 3.3   Calculating output node activations

		Input Activation	Connection Strength	Weighted Activations	Sum of Weighted Activations	Logistic of Net Input
Pattern 4	**Bias**	1	-1.756069	-1.756069	0.208519	**0.552**
	Input One	1	0.896755	0.896755		
	Input Two	1	1.067833	1.067833		

for each input pattern in Table 3.3. The actual output from the network is the number listed in the column entitled "**Logistic of Net Input**." These numbers match the output activations in Figure 3.15 exactly. Notice how the bias node effectively keeps the output node switched off (closer to 0 than 1) for the first 3 input patterns.

Exercise 3.7

1. Different random seeds result in different initial configurations of the connection weights in the network. So the presentation of training patterns to the network will result in different output activations for each random seed. This means that at the beginning of training different errors will be observed and this will, in turn, affect the way the weights are changed. In general, the error profile will be different for each random seed you train the network on. However, remember that each input pattern is trained on an associated target pattern. As the error on the output diminishes, the changes made to the weights in the network will diminish. The error then will change more slowly. In other words, different random seeds are likely to produce considerable variation in the error profile during the early stages of training but will show similar error profiles later in training when the error has been reduced. The error profile for running the network on two random seeds (1 and 6) are shown in Figure 3.22.

2. Although error generally decreases gradually during training, it need not do so in a monotonic fashion. For example, the error curve in Figure 3.22 (Seed 1) temporarily rises around the 400 sweep mark. This may seem odd given that the learning algorithm is continually attempting to reduce the error. There are several possible causes for U-shapes in our error curve. The less interesting reason is that the patterns most

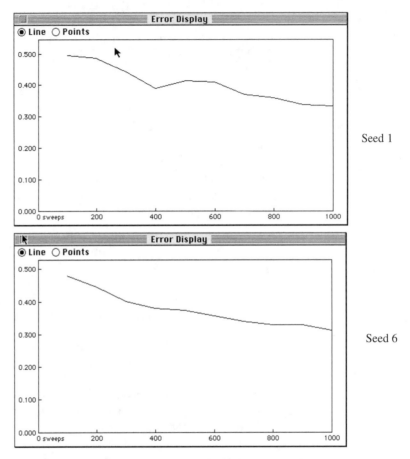

FIGURE 3.22   The error profiles for running Boolean AND on two different random seeds (1 and 6). Note that the final levels of error are very similar (about 0.35) but the trajectory of the error profiles are quite different.

recently sampled from the training set happen to have a large error associated with them. In other words, the reported error need not necessarily reflect the average RMS error for the 4 different training patterns—only a couple of them may have been selected for presentation to the network and those patterns may happen to have a large error.

A more interesting cause of the increase in the error is *interference* between different training patterns on the connection weights in the network. Consider a single input pattern. It is supposed to produce a specific output. There exists a configuration of connections in the network that will permit this. However, the network is also supposed to produce specific outputs for the other input patterns *using a single set of connections weights*. It is by no means obvious that the configuration of connections weights used for one pattern will be appropriate for the other patterns. If subsequent training patterns are not compatible with earlier ones, then further training will lead to changes in connections that mitigate against successful performance on the original pattern. Often the average error across all the patterns in the network will still decrease despite the increase in error on one of the patterns. However, sometimes the interference can be substantial and the average error may in fact increase.

3. Different networks can exhibit the same level of error for a given problem but need not arrive at those errors by the same route. In other words, the weights can look quite different even though they produce the same answer. For example, consider the weights files after 1000 sweeps of training for the two error curves in Figure 3.22. These are shown in Figure 3.23. The bias node has a stronger inhibitory effect in the "Seed

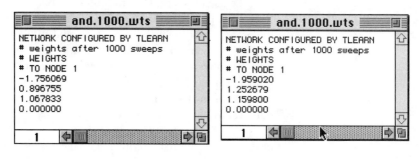

Seed 1   Seed 6

FIGURE 3.23   Two weights files for 1000 sweeps of training on Boolean AND. The final error is approximately the same but the weights are quite different. The bias node has a stronger inhibitory effect in the "Seed 6" simulation and the relative strengths of the connections from the input nodes to the output node are reversed.

6" simulation and the relative strengths of the connections from the input nodes to the output node are reversed. In general, there will be many solutions (configurations of the connection weights) to a problem. The solution that the network finds will depend on the initial start state, the number of times it sees a particular pattern and the order in which it sees different patterns.

4. Recall that the weight limit parameter in the `.cf` file determines the range of weight values that are randomly assigned to the connections in the network when it is initialized. If the weight limit is set to 4.0, then all the connections in the network will be set to values in the range ±2.0. In other words, some of the connections can be randomly set to be strongly excitatory (2.0) or strongly inhibitory (−2.0). We have already seen that the network's solution to Boolean AND is to build a strong inhibitory bias to the output node and moderate excitatory connections from the input nodes. If the initial weight assigned to the bias is strongly excitatory, the learning algorithm has more work to do to change it into a strong negative value. Similar difficulties will be encountered if the other two connections in the network have inappropriately large values. So a large weight limit has the potential to slow down learning in the network. Conversely, learning can sometimes be accelerated if the the bias node is born with a strong inhibitory connection to the output node. Time to find a solution would then depend on the value of the other connections in the network.

The use of a large weight limit can create other learning problems. Recall from Equation 1.5 on page 13 that the changes made to a weight are proportional to the δ value calculated for the associated target unit. δ itself is determined by the error on the output unit multiplied by the first derivative of the unit's activation function (see Equation 1.3). The first derivative of the activation function is simply the slope of the activation curve shown in Figure 1.3 on page 5. For large positive net input the output activation of a sigmoid unit is at its maximum, i.e., 1.0. However, the slope of the curve is flat. In other words, the value of the derivative is close to 0.0. Similarly, with large negative net input the slope of the curve is flat so its derivative is close to 0.0. When the derivatives are small, Equation 1.3 forces the δs to be small and so the weight changes will be small. In summary, extreme values of net input (positive or negative) lead to small weight changes. In effect, the units become saturated with input and find it difficult to learn. We will see later that this also has some beneficial side-effects.

Now a large weight limit can yield strong initial connections. Strong connections can yield extreme values of net input which lead to small weight changes. So large weight limits can slow down learning. Generally, connectionist modelers keep the initial weights small so they are not over-committed at the beginning of learning. A weight limit of 1.0, i.e., a range of ±0.5 seems to work quite well. This tends to keep the sigmoid units within their most sensitive range for learning.

Exercise 3.8

* Sometimes a higher learning rate will give you faster learning. The error curve with learning rate set to 0.5 in the Boolean AND problem is shown in Figure 3.24. In this case, learning benefits from a higher learn-

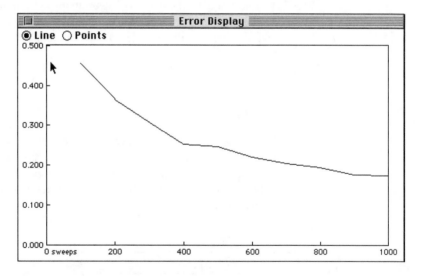

FIGURE 3.24   The error curve for Boolean AND with a random seed of 1 and a learning rate of 0.5. Learning is faster with the higher learning rate—compare with Figure 3.22.

ing rate. For example, the average error after 1000 sweeps is approximately 0.2. With a learning rate of 0.1 the average error after 1000 sweeps was closer to 0.35 (see Figure 3.22).

However, higher learning rate does not always lead to faster learning for all the patterns in the training set. We saw in Exercise 3.7 that input patterns in the training set can interfere with each other. If the learning rate is high, the likelihood of interference occurring is enhanced. The weight changes made for a single pattern presentation can have a detrimental effect on earlier training. Normally, it is safer to train the network with the learning rate parameter set to a small value unless there is good reason to believe that interference between patterns is likely to be minimal. Can you think why a higher learning rate seems to help in the Boolean AND problem?

Exercise 3.9

- You need a network that takes two inputs and produces a single output activation, i.e., two input nodes and a single output node. For good measure include a bias node. We will investigate its role in a later section.

Exercise 3.10

1. In Boolean AND the bias played an important role in keeping the output node switched off for all the input patterns except for 1 1. In Boolean OR the output unit should be switched on for all patterns except the first—0 0. So we would expect the bias to play a different role in the network for this problem.

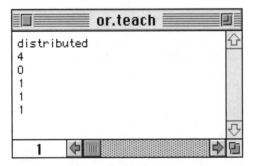

FIGURE 3.25   The **or.teach** file.

2. The **or.teach** file is shown in Figure 3.25. All output targets are set to 1 except the first (0) which codes the target for the input 0 0. The other **or** files are identical with their **and** counterparts since the input patterns and network architecture are identical.

Exercise 3.11

*   The weights file gives the best clue as to how the network has solved Boolean OR. **Open** the weights file **or.1000.wts** (or the most recently saved weights file) from the **File** menu. The weights file after 1000 sweeps with a random seed of 1 is shown in Figure 3.26. Notice

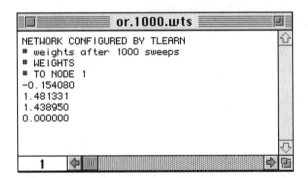

FIGURE 3.26   The **or.1000.wts** file. Note the small negative bias connection but relatively large positive connections from the input nodes.

that the connections from the input nodes are large and positive (1.48 and 1.44 respectively) which means that activity at either or both of the input nodes will tend to turn the output node on. The connection from the bias is negative but too small to counteract the contributions from active input nodes. In this example, the bias ensures that the output node is switched off when neither of the input nodes are active. Remember that the activation function for the output node varies continuously between 1 and 0 according to the logistic of the net input to the node (see Figure 1.3 on page 5). Net input of 0.0 to the output node produces an activation of 0.5 so a negative input is needed to switch off the output node when both input nodes are dormant (0 0). The mildly negative bias connection fulfils this function. However, the absolute size of the bias connection must remain smaller than either of the other two weights so that a single active input node can switch on the output node.

Exercise 3.12

- Given the network architecture that you have employed, the network will fail to solve XOR. Correct responses may be produced for some of the input patterns. With a random seed of 1 the network fails to produce correct responses for any of the patterns after 5000 sweeps of training. Try a variety of random seeds and learning rates. You will discover that the network manages to get some of the responses correct but *never* all of them at the same time.

CHAPTER 4   *Learning internal representations*

Introduction

In the previous chapter, you trained a single-layered perceptron on the problems AND and OR using the delta rule. This architecture was incapable of learning the problem XOR (see Table 3.1 on page 31). To train a network on XOR you need a multi-layered perceptron like the one shown in Figure 4.1. These networks contain a layer of hidden

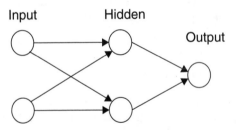

FIGURE 4.1 A simple multi-layered perceptron architecture for learning the Boolean function XOR

units that form "internal representations" of the input patterns. Connections propagate activity forward from the input nodes to the hidden nodes and from the hidden nodes to the output node. There are no connections between nodes within a layer or feedback connections to nodes in previous layers. Hence, network architectures of this type are often referred to as feedforward networks. During training, error is assigned to the hidden units by back propagating the error from the output. The error on the output and hidden units is used to change the weights on the second and first layer of connections respectively. Before continuing with the exercises in this chapter make sure that you are familiar with the exact procedure by which backpropagation adapts the weights in a network in response to an error signal. See "Learning" on page 10. You should also have completed Exercise 1.2 (c) on page 14.

Network configuration

Start up **tlearn** and **Open** a **New Project** called **xor. tlearn** opens three new windows called **xor.cf**, **xor.data** and **xor.teach**. Use the **xor.cf** file to create a network with the same architecture as that shown in Figure 4.1.

Exercise 4.1

- What are the contents of the three files?

Once you are confident that you have specified correctly all the necessary files (check your files with the answers to Exercise 4.1 on page 89) open the **Training Options...** dialogue box. Set the **Learning rate** parameter to 0.3 and the **Momentum** parameter to 0.9. Select a **Random seed** of 5 and set **tlearn** to **Train randomly** for 4000 sweeps.

Training the network

When you have specified all these training options open the **Error Display** and **Train the network**. If you have set all the training options as suggested, then you should observe an error display identical to that depicted in Figure 4.2. Check your training options if you do not observe this learning curve!

The error curve shows a sharp drop around the 2000 sweep mark. What has happened? First, let us examine the output activations for each of the input patterns after 4000 sweeps. The easy way to do this is to use the **Verify network has learned** option in the **Network** menu. The output from the network is shown in Figure 4.3. Again, if you do not observe these values in your own simulation, check that you have set the options in the **Testing options...** dia-

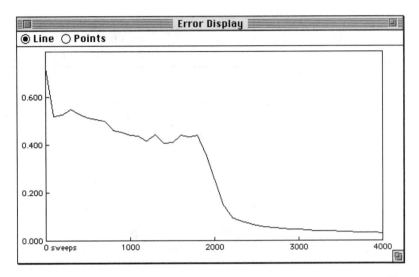

FIGURE 4.2 An error curve for a multi-layered perceptron trained on the XOR problem.

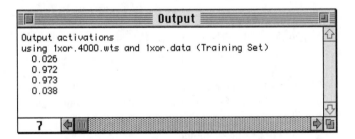

FIGURE 4.3 Output activations from multi-layered perceptron after training on Boolean XOR for 4000 sweeps.

logue box correctly (see Figure 3.20 on page 60 and accompanying text for instructions). Now consult the **xor.teach** file for the target activation values. The teacher signals for the four input patterns are **0**, **1**, **1**, **0** respectively. So the network has done very well at solving this problem. (Recall that we cannot expect the network to achieve exact matches to the target values—see Exercise 3.3 on page 63).

Exercise 4.2

- Can we tell from the error curve in Figure 4.2 if the
  network has solved XOR?

But why is there a sudden drop in error around the 2000 sweep mark?
To answer this question, it would be useful to examine output activations prior to this point in training. We can achieve this by saving
weight files during training of the network and then testing network
output with these weight files instead of the network's final (mature)
state.

Testing at different points in training

Open the **Training options...** dialogue box in the **Network**
menu. Notice the **more...** button in the bottom left-hand corner of
the dialogue box. Click on this button. The dialogue box expands to
offer a wider range of training options as shown in Figure 4.4. We

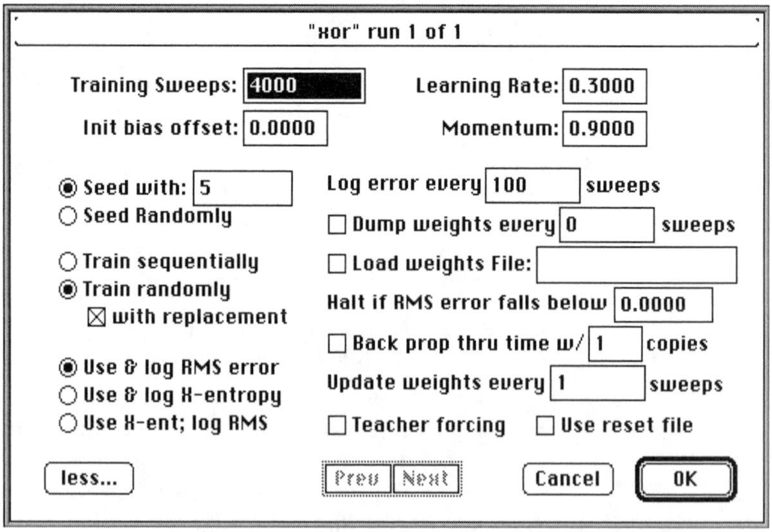

FIGURE 4.4   The **Training options...** dialogue box expanded to reveal the
full range of training options.

won't digress at this point to consider all the available options. For current purposes, the **Dump weights** option is of interest. Check this box and set the weights to be dumped every 500 sweeps. Leave all the other options as they are. Close the dialogue box and **Train the network** again. If you still have the **Error display** open, you should observe an identical error curve unfold. **tlearn** has performed an exact replication of the previous simulation using the same random seed (initial weight configuration) and same randomized order of pattern presentation. However, **tlearn** has saved weight files every 500 sweeps. Since the network has been trained for 4000 sweeps, 8 weight files will have been saved.

Next, open the **Testing options...** dialogue box and select the **Earlier one:** button. Click in the box to the right of the button to activate the dialogue box. Choose the **xor.1501.wts** file. Close the dialogue box and **Verify the network has learned**. **tlearn** will now display the output activations for the four input patterns after it has been trained for 1501 sweeps. By resetting the weights file in the **Testing Options...** dialogue box and **Verify**ing **the network has learned**, you can determine output activations at any point in training. Figure 4.5 shows the output activations after 1001,

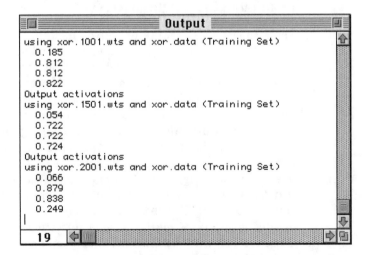

FIGURE 4.5   Sample output activations after 1001, 1501 and 2001 sweeps of training on XOR.

1501 and 2001 sweeps respectively. Using the rounding off criterion, we observe that the network has learned the first three patterns correctly already after 1001 sweeps. Overall, however, the network is behaving as though it was supposed to learn Boolean OR, i.e., it categorizes the fourth input pattern in the same fashion as the second and third input patterns. By the 1501 sweep mark the output activities on the first and fourth patterns have been reduced—a move in the right direction. However, the reduction in error on these two patterns has been achieved (apparently) at the expense of the second and third patterns. Notice that the output activities for the second and third patterns have reduced instead of increasing. The network is still behaving like Boolean OR. Finally, after 2001 sweeps of training the output activity for the fourth pattern has decreased considerably and the activity for the second and third patterns has increased again. Activity patterns have begun to move in the right direction and using the rounding off criterion, the network has now solved XOR. Notice that the period of training between 1501 and 2001 sweeps corresponds to the sudden drop in global error depicted in Figure 4.2. What has happened in the network to bring about this change?

Examining the weight matrix

The response of a network to a given input is determined entirely by the pattern of connections in the network and activation functions of the nodes in the network. We know what the activation functions are—**tlearn** specifies them in advance. However, the network organizes its own pattern of connections—the weight matrix. **tlearn** possesses a useful facility for examining weight matrices—the **Connection Weights** option in the **Display** menu. Open this display. As we saw in Figure 3.17 on page 52, **Connection Weights** displays a Hinton diagram of the weight matrix. In this case, we want to examine the weights at different points in training. Open the **Testing Options...** dialogue box and make sure that the **Earlier one:** button is set to **xor.2001.wts**—double click on the box to activate the dialogue box permitting you to select the **xor.2001.wts** file. Close the **Testing Options...** dialogue box and select the **Verify network has learned** option. **tlearn** will display the output activations again and update **Connection Weights** to display a

Hinton diagram for the state of the network after 2001 sweeps. You should obtain a Hinton diagram identical to that depicted in Figure 4.6.

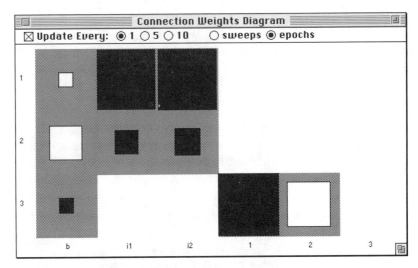

FIGURE 4.6 **Connection Weights** displays a Hinton diagram of the weight matrix for XOR after 2001 sweeps. Rows represent receiving nodes and columns represent sending nodes. White squares represent positive weights. Black squares stand for negative weights. The size of the square reflects the absolute value of the connection.

Recall that the Hinton diagram organizes connections by rows and columns. The row identifies the receiving unit and the column identifies the sending unit. So row 1 in Figure 4.6 identifies all the connections going into node 1 (the first hidden node). There are three connections—from the bias and the two input nodes. There are no feedback connections from the first hidden node to itself or from the second hidden node or output node to the first hidden node. Therefore, these areas in the Hinton diagram are left blank. We observe in Figure 4.6 that the first hidden node has a mildly positive connection from the bias node, so it is disposed to become active in the absence of any other input. In contrast, the first hidden node receives strong inhibitory connections from both input nodes. Thus, any activation coming up the input lines will tend to turn the first hidden node off.

Exercise 4.3

- Draw a diagram of the network after 2001 sweeps depicting all connections (including the bias) as positive or negative.

In a later section we will examine how this network connectivity manages to provide a solution to the XOR problem. For the time being however we are concerned with the *changes* that occur in the network between 1501 and 2001 sweeps that enable the network to solve the problem. Now examine the weight matrix after 1501 sweeps of training. Open the **Testing Options...** dialogue box and set the **Earlier one:** option to **xor.1501.wts**. Close the dialogue box and **Uerify the Network has learned**. The **Connection Weights Diagram** will be updated to depict a Hinton diagram for the weight matrix at 1501 sweeps as shown in Figure 4.7.

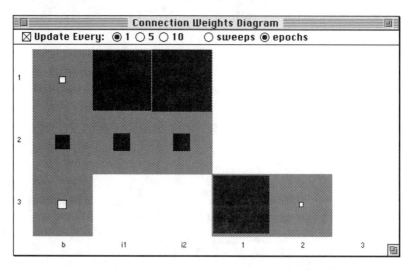

FIGURE 4.7   Connection Weights displays a Hinton diagram of the weight matrix for XOR after 1501 sweeps.

How does the weight matrix shown in Figure 4.7 change into the weight matrix shown in Figure 4.6? The most significant changes take place in the connections from the bias node to the hidden and output nodes, and from the second hidden node (node 2) to the output node

(node 3). The remaining connections do not change much. In particular, the bias connections to the hidden nodes grow in strength to provide positive excitation to their target nodes while the bias connection to the output node switches to inhibit its activation. The connection from the second hidden node to the output node increases its capacity to excite the output node. These changes take place in unison, as they must if they are to permit the fourth input pattern to be classified correctly.

Exercise 4.4

- Draw a diagram of the network depicting the exact values of the connections (to one decimal place) at 1501 and 2001 sweeps. Can you see how the network fails to solve the XOR at 1501 sweeps but passes the test at 2001 sweeps? *Hint: You will need to examine the weight files themselves as shown in Figure 3.26 on page 73.*

Hinton diagrams provide a convenient overview of the connectivity in the network. You can even request that **tlearn** displays changes in the weight matrix on-line. Try it. Make sure that the **Connection Weights** diagram is displayed and then simply **Train the Network**. Initially, you will observe substantial swings in the weight values in the network. However, weight values will gradually stabilize. If you have the **Error Display** active then you can also observe how changes in the global RMS error coincide with changes in the weight matrix. The next stage in analyzing network performance involves examining hidden node activations.

Hidden node representations

The activations of the hidden nodes provide an important clue as to how the network has solved XOR. In general, we may consider the hidden unit activations associated with an input pattern to be the network's *internal representation* of that pattern. We shall discover that patterns that look quite dissimilar at the input layer can be almost identical when we view their activations at the hidden unit level. Conversely, patterns that are similar at the input layer can end up looking quite different at the hidden layer. Hidden units and the connections feeding into them have the capacity to transform the similarity relations of the patterns in the input space so that the nature of the problem itself is changed.

We shall investigate several ways of examining these activities in the network. To begin with open the **Testing Options...** dialogue box and set the **Earlier one:** option to `xor.2001.wts`. Close the dialogue box and open the **Node Activation** display in the **Display** menu. Click on the (**First Data Pattern**) button. The display should update to reveal the display in Figure 4.8. This represents the

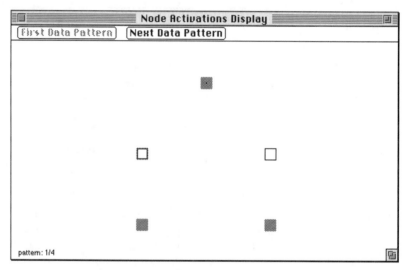

FIGURE 4.8   Node Activation Display depicts the activation of all nodes in the network (excluding the bias). The level of activation of each node determines the size of the white square. A grey node indicates an inactive node. In the current display, the input and output nodes are inactive and the hidden nodes are active.

pattern of activity in the network when the first input pattern 0 0 is presented to the network. The current pattern is identified in the bottom left-hand corner of the display. All the nodes in the network are represented (except for the bias node). The level of activation of a node is represented by the size of the white square—large squares reflect high activity. Grey squares indicate dormant nodes. Input nodes are displayed at the bottom of the diagram. Subsequent layers in the network are depicted at higher levels in the display. Hence, in Figure 4.8, the input nodes are dormant, the output node is dormant (as it should be for correct performance) and the hidden nodes are active. By clicking on the [Next Data Pattern] button, you can display node responses to all of the four input patterns.

Exercise 4.5

- Examine the activation of the hidden nodes in response to the different input patterns. Can you see how the network is solving the non-linear XOR problem? Why is it that the first input pattern produces highly active hidden nodes but a dormant output node?

Just as it is possible to determine the exact values of the connection weights in the network, it is also possible to determine the exact activations of the hidden nodes. Open the **xor.cf** file using the **Open** option from the **File** menu (if it isn't already open). Observe that in the **SPECIAL:** section there is a line with the instructions: **selected = 1-3**. This line tells **tlearn** to output the activations of the selected nodes whenever the **Probe Selected Nodes** option from the **Network** menu is chosen. Make sure that the **Testing Options...** is still set to use the weights file **xor.2001.wts** and then select **Probe Selected Nodes**. **tlearn** will display activations for nodes 1–3 for the four input patterns as shown in Figure 4.9.

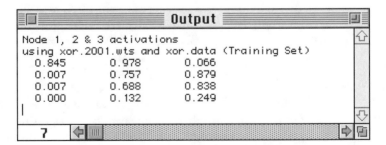

FIGURE 4.9   Node activations in the **Output** window for nodes 1–3 in a network trained on XOR for 2001 sweeps. The output is produced by selecting the **Probe Selected Nodes** option in the **Network** menu. Nodes are selected in the **SPECIAL:** section of the **xor.cf** file.

Notice that the third column of activation values is identical with the activations displayed in Figure 4.5 for 2001 sweeps. The first two columns in Figure 4.9 list the activations of the two hidden nodes for the four input patterns. These values give the precise values used to calculate the **Node Activations** display in Figure 4.8.

Exercise 4.6

---

1. Use the weight matrix from **xor.2001.wts** (see Exercise 4.4) to confirm that the activation values listed in Figure 4.9 are correct. *Hint: You will need to refer to Exercise 1.1 on page 8 to convert the net input to a node into its activation value.*

2. Have we explained the *sudden* decrease in the global error score observed in Figure 4.2?

Role of learning rate and momentum

The learning rate parameter η (see Equation 1.4 on page 12) determines what proportion of the error derivative is used to make changes to the weights in the network. In the **Training Options...** dialogue box you are also given the option of specifying the value of another parameter called momentum (μ). The momentum parameter provides another means for manipulating the weights but based on the changes that were made on the previous learning trial. The value of the momentum parameter determines what proportion of the weight changes from the previous learning trial will be used on the current learning trial. Mathematically, we can express this notion as

$$\Delta w_t = -\eta \frac{\partial E}{\partial w} + \mu \Delta w_{t-1} \qquad \text{(EQ 4.1)}$$

which reads: "The change in the weight at time t is equal to the learning rate η multiplied by the negative slope of the error gradient with respect to the weight, plus the momentum μ multiplied by the size of the weight change on the previous learning trial." Thus, if the weight change on the previous learning trial is large and μ is large, the weight change on the current learning trial will also be large even if the derivative of the error with respect to the weight (i.e., $\partial E / \partial w$) is small. In other words, the momentum term has the effect of maintaining weight change in the network when the error surface flattens out and speeding up weight changes when the error surface is steep. Of course, if $\mu = 0$ then the learning algorithm operates in the usual fashion.

Exercise 4.7

---

1. It has been suggested that momentum can help the learning algorithm solve problems like XOR which would otherwise be very difficult to solve. Do you agree with this suggestion? Under what conditions might it be true?

2. Try running the XOR simulations again but experiment with different values of learning rate and momentum. Can you find an optimum combination of learning rate and momentum for the XOR problem? In general, is it best to have high or low levels of learning rate and momentum for XOR? When you have found another combination of learning rate and momentum that solves XOR, analyze the solution the network has found and compare it with the solution achieved in this chapter. Is there only one solution to XOR?

Role of hidden nodes

You have already discovered the importance of hidden nodes for solving the XOR problem. Now examine the different ways in which the quantity and organization of hidden nodes can effect network performance.

Batch versus pattern update

Examine the expanded version of the **Training Options...** dialogue box. You will notice that there is an option called **Update weights every:** which has been set to 1 sweep. This option tells **tlearn** when to make weight changes in the network. Until now, we have used the default setting for this option which is to update the weights after

Exercise 4.8

1. Reconfigure your network to use different numbers of nodes in the hidden layer, say from 1 to 5. Does increasing the number of hidden units assist the network in solving the task? Determine the role of the individual units in solving the task.

2. Configure the network with an additional layer of 2 hidden nodes. Does the additional layer accelerate learning on XOR? Examine the internal representations constructed in the different hidden layers.

every pattern presentation. In network jargon, this is called pattern update. Of course, it is possible to delay making weight changes until the network has seen the other training patterns. A common training regime, called batch update, involves presenting all input patterns just once, accumulating the calculated changes for each connection and then updating the weights for all pattern presentations simultaneously. In the XOR problem, this is equivalent to setting **Update weights every:** to 4 sweeps.

Exercise 4.9

- What do you think are the consequences for the learning profile of the network when choosing between pattern and batch update? Can you think of any computational advantages in using batch mode? Test your theory by running XOR in batch mode. *Note: You will also need to deactivate the* **With Replacement** *under the* **Train randomly** *check box if you wanted to guarantee that the network sees each training pattern on every epoch. Of course, direct comparison with a pattern update training schedule would then be inconclusive since until now you have selected patterns randomly with replacement.*

Answers to exercises

Exercise 4.1

- The **xor.cf** file should contain the following information:

```
NODES:
nodes = 3
inputs = 2
outputs = 1
output node is 3
CONNECTIONS:
groups = 0
1-2 from i1-i2
3 from 1-2
1-3 from 0
SPECIAL:
selected = 1-3
weight_limit = 1.00
```

The **NODES:** section indicates that there are 3 nodes in the network—2 hidden nodes and 1 output node (input nodes don't count). The **CONNECTIONS:** section contains an additional line for the extra layer of connections in the multi-layered perceptron. There is also a connection from the bias to the hidden nodes and the output node (**1-3 from 0**). Notice how it is possible to specify network architectures of increased complexity through minor additions to the **.cf** file.

The **xor.teach** file requires only a minor change in relation to the **or.teach** file:

```
distributed
4
0
1
1
0
```

The **xor.data** file is identical to the **or.data** file.

Exercise 4.2

- Whether the network has solved the problem depends on the error criterion you decide to use. If you adopt the "rounding off" procedure introduced in Exercise 3.3 on page 63 and further discussed in Exercise 3.4 on page 65 then a level of global RMS error, namely 0.25 , can be used to *guarantee* successful performance on all four input patterns. The global RMS error displayed in Figure 4.2 falls below this value around the 2000 sweep mark, so the error curve alone tells us that the network has solved the problem. Of course, the network may have solved the problem before this point in training.

Exercise 4.3

- Figure 4.12 shows the connectivity of the network after 2001 sweeps of training on the XOR probelm. Each connection is depicted as inhibitory (-) or excitatory (+).

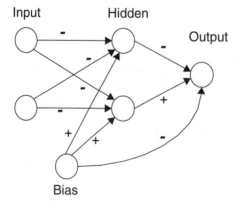

FIGURE 4.10   Connections weights in multi-layered perceptron trained for 2001 sweeps on XOR.

Exercise 4.4

- Use the weight files **xor.1501.wts** and **xor.2001.wts** to obtain the exact values of the connections in the network. You can **Open** these files from the **File** menu. Their contents are shown in Figure 4.11. Con-

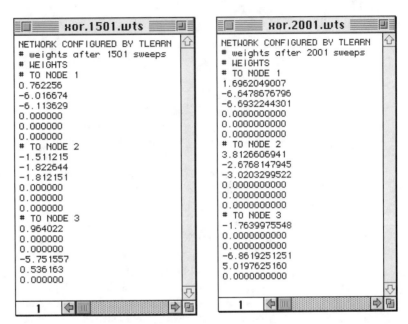

FIGURE 4.11   Weight files after 1501 and 2001 sweeps of training on XOR.

nections are listed for each target node. The connection from the bias is listed first, then connections from the input units and so on up through the network. A value of 0.000000 will almost always indicate that the connection is non-existent. The state of the network after 2001 sweeps of training is shown in Figure 4.12.

Exercise 4.5

- Both hidden units are highly active for the first input pattern (0 0) while only the second hidden unit is active for the second and third input patterns (1 0) and (0 1). Neither of the hidden units are active for the last

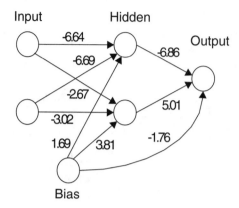

Input   Hidden

-6.64   Output

-6.69

-6.86

-2.67

-3.02   5.01

1.69   3.81   -1.76

Bias

FIGURE 4.12   Connections weights in multi-layered perceptron trained for
2001 sweeps on XOR.

input pattern (1 1). In other words, the internal representations of the
second and third input patterns are more or less identical and so will produce the same output activation (1)—as they should. In this case, the
excitatory activity from the second hidden node is sufficient to overcome
the negative effect of the bias and turn on the output unit. For the first
and fourth input patterns, the negative connections from the bias and the
first hidden unit guarantee that the output unit remains switched off. The
layer of hidden units have transformed a *linearly inseparable* problem
into a *linearly separable* one (see Chapter 2 of the companion volume,
Rethinking Innateness, for a discussion of *linear separability*).

Exercise 4.6

1.   Repeat the procedure you performed in Exercise 3.6 on page 66. Note,
     however, that in this case you will also need to use the activations of the
     hidden units to calculate the activations of the output unit for the 4 input
     patterns. Don't forget to include the bias in your calculations!

2.   Although we have extracted a good deal of information from **tlearn**
     as to how the network solves XOR, we have not explained why there is a
     sudden decrease in the global error between 1501 and 2001 sweeps.
     There are several ways to understand how the network suddenly finds a
     solution to the XOR problem.

The learning algorithm that we use in this book, backpropagation, belongs to a family of *gradient descent* learning algorithms. The idea is that the learning algorithm attempts to decrease the error at the output by traversing an *error landscape* in a downhill direction. It achieves this by calculating the slope of the error surface at the current location and making changes in the weights which will take it downhill. This is the calculation embodied in in the partial derivative introduced in Equation 1.4 on page 12. The idea of gradient descent is discussed in more detail in the companion volume *Rethinking Innateness*, Chapter 2 (page 71). The reason why the network suddenly finds a solution to XOR is that the error surface has a sharp dip in it, such that a small change in the value of the weights brings about a relatively large change in the output error.

Another way to understand how the network suddenly finds a solution to XOR requires that you understand the idea of *linear separability* (also discussed in more detail in *Rethinking Innateness*, Chapter 2 (page 62–63). We'll review the problem briefly again here. It is possible to visualize the Boolean functions AND, OR and XOR graphically in a two dimensional plane as shown in Figure 4.13. Different partitionings of the

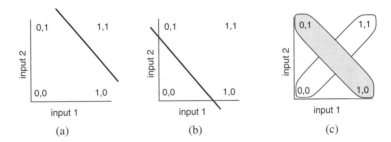

FIGURE 4.13   Geometric representation of the XOR problem. If the four input patterns are represented as vectors in a two-dimensional space, the problem is to find a set of weights which implements a linear decision boundary for the output unit. In (a), the boundary implements logical AND. In (b), it implements logical OR. There is no linear function which will simultaneously place 00 and 11 on one side of the boundary, and 01 and 10 on the other, as required for (c).

space correspond to different Boolean functions. The partitionings shown in Figure 4.13 (a) and (b) represent Boolean AND and OR, respectively. Notice how the space can be partitioned appropriately simply by drawing a line separating three patterns from the other one. The problems are linearly separable. In fact, there are an infinite number of lines that you could draw that would partition the space appropriately for these two problems (different angles and positions). Each of these lines corresponds to a different solution that the network might find, i.e., a different configuration of the weight matrix. In contrast, XOR is linearly non-separable—it is not possible to partition the space with a single line such that all patterns are placed in the correct partition. This is why you

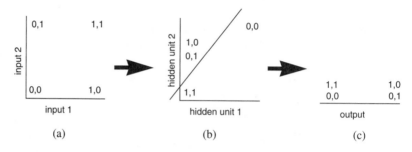

FIGURE 4.14   Transformation in representation of the four input patterns for XOR. In (a) the similarity structure (spatial position) of the inputs is determined by the form of the inputs. In (b) the hidden units "fold" this representational space such that two of the originally dissimilar inputs (0,1 and 1,0) are now close in the hidden unit space; this makes it possible to linearly separate the two categories. In (c) we see the output unit's final categorization (because there is a single output unit, the inputs are represented in a one-dimensional space). Arrows from (a) to (b) and from (b) to (c) represent the transformation effected by input-to-hidden weights, and hidden-to-output weights, respectively.

need a network with hidden units to solve the problem. As we saw in Exercise 4.5 on page 91, the hidden units transform a linearly non-separable problem into a linearly separable one. However, the hidden units cannot do this immediately. The connections between the input units and the hidden units (and the bias) have to be adapted to an appropriate configuration by the learning algorithm. The configuration is shown in Figure 4.12.

Think about this again from a two dimensional point of view. We saw in Exercise 4.5 that the network treats the input patterns (1 0) and (0 1) as more or less identical at the hidden unit level. This is represented graphically in Figure 4.14. Once the hidden unit activations corresponding to the four input patterns become linearly separable in accordance with the problem at hand, the network can find a solution. This corresponds to the hidden unit activation of one of the input patterns (in this case 1 0) moving into a region of the space occupied by its partner (in this case 0 1). A line can then partition the hidden unit activation patterns into their appropriate categories and produce the correct output activations. Before this point, no solution is possible and the error remains high.

For example, after 1501 sweeps of training in the network, the activations of the hidden and output nodes can be probed to yield the activation values shown in the **Output** window shown in Figure 4.15. The

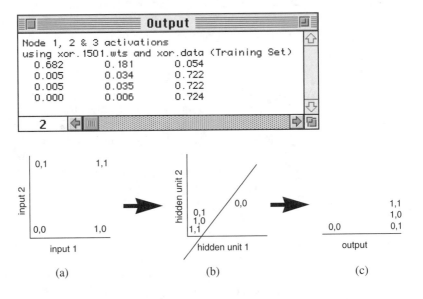

FIGURE 4.15 Transformation in representation of the four input patterns for XOR after 1501 sweeps of training. The network is responding as if it had been trained on Boolean OR. The network achieves the solution depicted in Figure 4.14 by increasing the activation of the second hidden unit for input patterns (1 0) and (0 1) and shifting the demarcation of the linear partition.

corresponding positions of the patterns in two dimensional space are also shown in Figure 4.15. Clearly, the hidden unit activations do not permit a linear separation of the problems. In fact, the network is behaving as though it had been trained to solve Boolean OR. To solve XOR the second hidden unit activations for the patterns (1 0) and (0 1) must be higher. Once the network has achieved this, the problem becomes linearly separable and the solution is quickly reached.

Exercise 4.7

1. Generally, momentum helps when the patterns in the training set have to be treated in similar ways. Learning on previous patterns can be transferred to subsequent patterns. Sometimes, momentum can help the network avoid getting trapped in local minima—local regions on the error landscape where the error gradient is zero (see *Rethinking Innateness*, Chapter 2, page 71).

2. The identification of an optimum combination of learning rate and momentum depends on the start state of the network. Since this is often determined in a random fashion, it is difficult to specify what the appropriate values of these parameters should be for any given simulation. However, it is generally a good idea to use a small learning rate to avoid situations where new learning wipes out old learning.

    You will find that there are many solutions to XOR. Some of them are quite surprising!

Exercise 4.8

1. You will find that the network seems to do best with just 2 or 3 hidden units. Although there are many solutions available to the network with 5 hidden units, a common solution is to turn off the hidden units that the network doesn't need by developing strong inhibitory connections to them from the bias node. In this manner, the network comes to behave as if it only had 2 or 3 hidden units.

2. Adding extra layers of hidden units to the network rarely speeds up the time to solution. On the contrary, the extra layers of randomized weights obstructs the construction of good internal representations since the backpropagation of error (assignment of blame) is reduced from one layer to the next (see Equation 1.6 on page 14).

Exercise 4.9

- Since weight changes made in batch update reflect the *average* error for all the patterns in the training set, learning is likely to follow a smoother curve than pattern update. Training will also be faster in batch mode because fewer weight changes are being made on each epoch of training.

CHAPTER 5 # Autoassociation

Introduction

Feedforward networks are trained to approximate a function associating a set of input patterns with a set of output patterns. It is also possible to teach networks simply to reproduce their input. In this case, the input and output layers have the same number of elements. The network is given a pattern and its task is to pass that pattern through itself until it is reproduced on the output layer. Thus, the input and the teacher pattern are identical. In such circumstances, the network is said to be doing *autoassociation* or *identity mapping*. In this chapter, you will:

- Learn how autoassociation can be used to convert a "local" representation into a "distributed" representation. This is sometimes called the *encoding* problem.

- Investigate how autoassociation can also be useful in discovering redundancies in a set of input patterns which lead to more efficient featural representations.

- Show how autoassociators can perform efficient pattern completion in the face of noisy input.

- Develop additional techniques for analyzing hidden node representations.

Local and distributed representations

In this section, you will encode 4 bit vectors. Therefore, you will need a network containing 4 input nodes and 4 output nodes. Create a **New Project...** called **auto1**. Configure the **auto1.cf** with 2 hidden nodes, such that the 4 input nodes are all connected to the 2 hidden nodes and the 2 hidden nodes are all connected to the 4 output nodes in a strictly feedforward fashion. Connect a bias to the hidden and output nodes. The weights should be randomized within the range ±0.5.

Each pattern consists of a 4-bit vector (each bit will be either a 1 or a 0). Furthermore, the patterns you will use in this example are what might be called "local representations." These are the patterns:

```
1 0 0 0
0 1 0 0
0 0 1 0
0 0 0 1
```

Exercise 5.1

- What is meant by "local representation?" In what sense are the above patterns instances of local representations?

Create an **auto1.data** file and **auto1.teach** file appropriate for the task. Note that **tlearn** offers an alternative format for data presentation when the patterns are represented in a localist fashion, i.e., when only a few of many input nodes are non-zero. In the "localist" format, the input patterns in the **.data** file are a list of nodes, specifying only the numbers of those nodes whose values are to be set to one. The node specification follows the conventions used in the **.cf** file. See "The Configuration (.cf) file" on page 38. All other input nodes are assumed to be zero. Localist coding of the output patterns in the **.teach** file can be used to specify the output nodes which are to be on. In other words, an output pattern in the localist case is a set of integers designating the nodes whose target values are one. All other nodes are assumed to be zero. Possible **data** and **teach** files are shown in Figure 5.1. You may continue to use the "distributed"

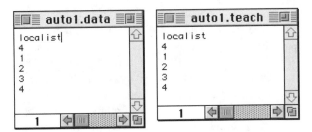

FIGURE 5.1   The `auto1.data` and `auto1.teach` files using localist coding
mode.

mode of data coding if you wish. It is introduced here purely as a
notational variant.

Check that you have set up your configuration file correctly by
displaying the **Network Architecture**. `tlearn` should depict a
network with the same pattern of connectivity as displayed in
Figure 5.2. If you do not obtain this display, then check your
`auto1.cf` file.

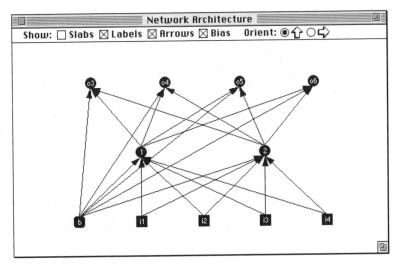

FIGURE 5.2   Network Architecture for `auto1`.

Training the network

Feel free to experiment with the range of training options available to you. For purposes of exposition, however, we will refer to a simulation in which the following training options have been selected: Set the learning rate parameter to 0.3 and specify a momentum of 0.9. Specify a random seed of 1 and train randomly, updating the weights after every sweep (pattern update). Train the network for 3000 sweeps. You may also wish to activate the error display to examine the global RMS error.

Exercise 5.2

- What level of global error guarantees that **tlearn** has found a solution to the encoding problem? (Assume the rounding off criterion for evaluation purposes.)

When you have found a configuration of training parameters that yields a solution to the encoding problem, select the **Node Activation** display. Make sure that you have selected the **Most recent** weights file in the **Testing Options...** dialogue box. You should observe a display like that shown in Figure 5.3. Cycle through the four input patterns and check that the network is autoassociating in the required fashion. Cycle through the four input patterns again and see how the hidden units have encoded these 4 patterns. The hidden unit activations will have learned to encode their input/output pattern pairs by using a distributed representation of the patterns.

Exercise 5.3

---

1. What is the key to this distributed representation? Why is the representation "distributed"?

2. Open up the weights file **auto1.3000.wts** and look at the network which did the encoding. Draw the network. Show weights on each of the connections, and put the biases inside the nodes.

You have seen how a network can represent 4 input patterns using 2 bits. What do you think would happen if you modified the network so

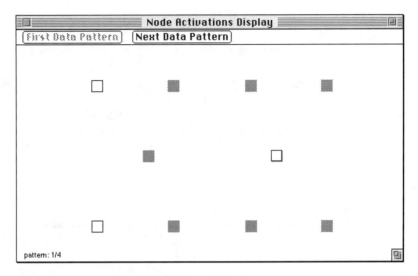

FIGURE 5.3   **Node Activations Display** for **auto1** after 3000 training sweeps.

that it had 5 inputs and 5 outputs, but still only 2 hidden units; and then trained it to encode the following patterns?

```
1 0 0 0 0
0 1 0 0 0
0 0 1 0 0
0 0 0 1 0
0 0 0 0 1
```

Exercise 5.4

- Do you think the network could do the task? Construct such a network and report your results. So as not to override the work on **auto1**, call this project **auto2**.

Pattern completion

Now return to the original autoassociation task in this exercise—
auto1. Modify your **auto1.data** file so that one of the input patterns is distorted. For example, you might change the input pattern
1 0 0 0 to **0.6 0 0 0.2**. You've just introduced noise into the
input stream. In the **Testing Options...** dialogue box select the
auto1.3000.wts file which you created earlier. Now test your network again using the **Verify Network has learned** option.
Alternatively, you can observe the response on the **Node Activation Display**.

Exercise 5.5

> • Has the network partially reconstructed the noisy
>   pattern? Experiment with input patterns that have
>   noise in different locations and in differing amounts.
>   Is there some critical noise level (or combination of
>   noise values) for which the pattern completion fails?

This phenomenon, called *pattern completion*, is a result of the com-
pressed internal representation. The difference between a degraded
and a pristine input pattern is partially eliminated when the activation
goes through the nonlinearity at the hidden layer.

Feature discovery

In this section, you will investigate another purpose to autoassociation, which is to discover a featural representation of the input.
Because the intermediate (hidden) layer(s) are generally of smaller
dimensionality than the input and output layers, for the autoassociation to work, the network is required to funnel the pattern of input
activation through its narrow "waist." The pattern of hidden unit activations must be efficient and concise; if the hidden units are to successfully regenerate the input pattern on the output layer, they must
discover some representation of the input which is reduced but which
contains all the information necessary to recreate that input. This task

encourages the network to discover any structure present in the input so that the input can be represented more abstractly and compactly. The network need know nothing about the significance of the input.

Exercise 5.6

• What does it mean to say that the concept of "redundancy" is a property of a *set of patterns*?

To do this exercise, you will need a network that has 7 input nodes and 7 output nodes, and 3 hidden nodes. The **data** file should contain the following 8 input patterns:

```
1 1 0 1 0 0 1
1 0 0 1 1 1 1
1 0 1 1 1 0 0
1 1 0 1 0 0 0
0 1 0 0 0 0 1
0 1 0 0 0 1 0
0 0 0 0 1 0 0
0 0 1 0 1 1 0
```

Since the task is autoassociation, the input and output patterns will be identical. You will need to study the activation of the hidden nodes so make sure that they are **selected** in the **SPECIAL:** section of the **.cf** file.

Exercise 5.7

---

1. Before running the network, study the input patterns. Do you see any way(s) of grouping the patterns? That is, do any of the patterns resemble each other? Are there different ways of grouping the patterns?

Exercise 5.7

---

2. Train the network for 4000 sweeps. (The results reported below use a random seed of 1, learning rate of 0.3 and momentum of 0.9. However, you may choose a different configuration if you wish. Just remember that in this case the results will look a bit different.)

3. After the network is trained, test the network to see that it has learned to generate the correct output. Has the network learned to regenerate the input patterns? How well?

Now see if there is any similarity structure to these patterns. This involves looking at the hidden node activation patterns. Do this twice. First, test the activations of the 3 hidden nodes in response to each of the 8 probes. You can display hidden node activations by selecting the **Probe Selected Nodes** option in the **Network** menu. If you have specified the hidden nodes as "selected" in the **SPECIAL:** section of the **.cf** file, then the hidden node activations will be displayed in the **Output** window as shown in Figure 5.4. Can you discern any patterns? Probably not; so do something else. This involves editing the **Output** window containing the activation of the hidden nodes First, clear the two lines at the beginning of the file containing header information so that you are left with a file containing 8 rows and 3

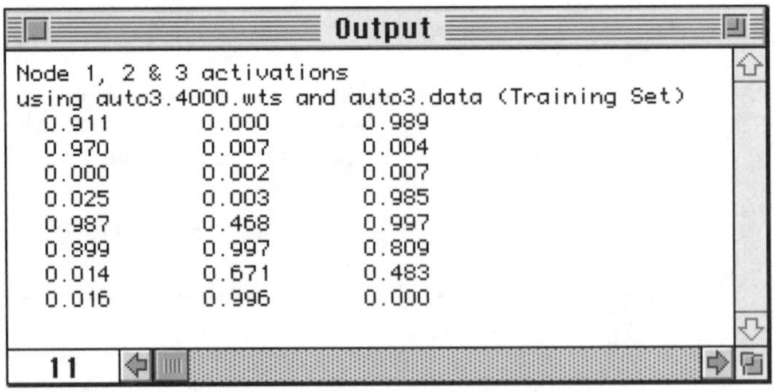

```
                              Output
 Node 1, 2 & 3 activations
 using auto3.4000.wts and auto3.data (Training Set)
     0.911          0.000          0.989
     0.970          0.007          0.004
     0.000          0.002          0.007
     0.025          0.003          0.985
     0.987          0.468          0.997
     0.899          0.997          0.809
     0.014          0.671          0.483
     0.016          0.996          0.000

    11
```

FIGURE 5.4   Hidden node activations in an autoassociation task produced by **Probe Selected Nodes** in the **Network** menu.

columns, each row specifying the hidden node activations for each input pattern. Insert a name representing the input pattern at the beginning of each line. You can use any numeric characters you choose for this label just so long as it doesn't contain a space. When you have edited the **Output** window it should resemble the window displayed in Figure 5.5. Use the **Saue as...** option in the **File** menu to save this file as **auto3.sort**.

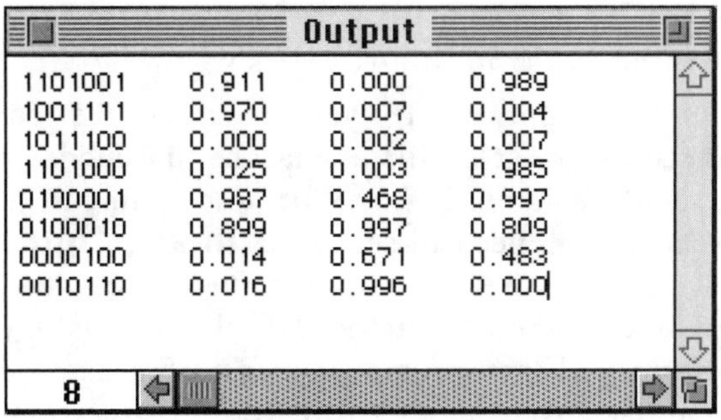

Output			
1101001	0.911	0.000	0.989
1001111	0.970	0.007	0.004
1011100	0.000	0.002	0.007
1101000	0.025	0.003	0.985
0100001	0.987	0.468	0.997
0100010	0.899	0.997	0.809
0000100	0.014	0.671	0.483
0010110	0.016	0.996	0.000

| 8 |

FIGURE 5.5   Hidden node activations listed by pattern name.

Your goal is to see whether the internal representations (that is, the hidden node activation patterns) give you any clue as to the similarity structure of the patterns you have autoassociated. To do this, it would be useful to be able to group the patterns in various ways. You might wish to sort all the patterns according to the values of the first hidden node, for instance. This would group (or classify) the patterns according to how the hidden node "interpreted" them. It might be that the first hidden node discovered some feature which is present in some of the patterns but not in others, so that you find two groups of patterns. (Or perhaps not.) Similarly, you would also like to sort the patterns according to the values of the second hidden node, and the third hidden node. You can do this easily using the **Sort...** utility. This utility will sort a file, row by row. You can indicate where you want to start in the row and skip over certain columns.[1] Select the **Sort...** option from the **Edit** menu. The dialogue box depicted in

---

1. You can even perform a nested sort whereby rows are first sorted by one column. Rows which match on one column then undergo sorting in a second (user specified) column.

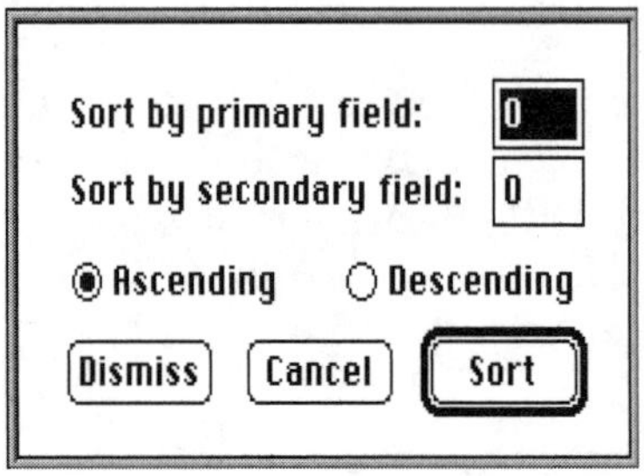

FIGURE 5.6 The **Sort...** dialogue box.

Figure 5.6 appears. If you look at the format of your file **auto3.sort**, you will see that the first hidden node activation is in the second column. Thus, to **sort** according to values of the first hidden node, you need to set the **primary field:** box in the **Sort...** dialogue to 1. We are not concerned here with performing a nested sort so leave the **secondary field:** box as 0. Click on [OK]. **tlearn** will then prompt you for a file name in which to save the sorted data. You can then **Open...** this file from the **File** menu. The lines in the resulting file will be displayed in a different order to those in **auto3.sort**. You will see that they are ordered by ascending values of the first hidden node. In a similar way, you can sort by the remaining hidden node activations. Just select the appropriate **Primary field:** in the **Sort...** dialogue.

Exercise 5.8

1. You will probably find at least one, and probably two, hidden nodes which nicely classify the input patterns. What aspects of the patterns does each hidden node seem to be attending to? Are the features in the patterns confined to single bits, or multiple bits? Do all hidden nodes succeed in finding features?

2. Earlier we said that redundancy was a property of sets of patterns. What implications does this have for a network which was taught on a random subset of 3 of the above patterns. Would you expect similar hidden node patterns to develop? Why or why not?

Exercise 5.8

3. Let's say that you believe you have found a feature in the patterns which is being extracted by the network (extracted in the sense that a hidden node always is highly activated whenever a pattern with that feature is present). It would be interesting to be able to test your hypothesis, using the network you have just trained. How might you test your hypothesis? Do the test and report your results.

Answers to exercises

Exercise 5.1

- Localist representations are similar in some ways to traditional symbolic representations. Each concept is represented by a single node, which is atomic (that is, it cannot be decomposed into smaller representations). The node's semantics are assigned by the modeler and are reflected in the way the node is connected to other nodes. We can think of such nodes as hypothesis detectors. Each node's activation strength can be taken as an indicator of the strength of the concept being represented.

    The advantage of localist representations is that they provide a straightforward mechanism for capturing the possibility that a system may be able to simultaneously entertain multiple propositions, each with different strength, and that the process of resolving uncertainty may be thought of as a constraint satisfaction problem in which many different pieces of information interact. Localist representations are also useful when the modeler has a priori knowledge about a system and wishes to design the model to reflect that knowledge in a straightforward way. Finally, the one-node/one-concept principle makes it relatively easy to analyze the behavior of models which employ localist representations.

    The input patterns used in this exercise are localist because just one node tells you all you need to know about to identify the pattern from all the others in the training set.

Exercise 5.2

- Using the same logic as in Exercise 3.3 on page 63, the maximum RMS error that guarantees that all patterns have been properly encoded is given by the expression:

$$\text{RMS error} \leq \sqrt{\frac{(0.5^2)}{4}} = 0.25$$

    Notice that this result is identical to the answer to Exercise 3.3. In both cases there are 4 input patterns and only one output node is supposed to be active.

Exercise 5.3

1. If you used the same network parameters as we suggested, you should find that the input patterns produce the hidden unit activations shown in Table 5.1. Different training parameters will yield different representa-

TABLE 5.1 Hidden unit representations of four localist input patterns in an autoassociator.

Activations	
Input	Hidden
1 0 0 0	0 1
0 1 0 0	0 0
0 0 1 0	1 1
0 0 0 1	1 0

tions at the hidden unit level. The solution shown in Table 5.1 is particularly clear: Each input pattern produces a unique binary pattern across the hidden nodes. The two-dimensional hidden unit vector makes optimal use of the space of activations available to it (cf. Figure 4.14a). The different hidden unit activations permit the network to distinguish the input patterns and reproduce them at the output.

In a distributed representation, a common set of units is involved in representing many different concepts. Concepts are associated, not with individual units, but instead with the global pattern of activations across the entire ensemble. Thus, in order to know which concept is being considered at a point in time, one needs to examine the entire pattern of activation (since any single unit might have the same activation value when it participates in different patterns). For example, the activation level of the first hidden unit in **auto1** is insufficient to identify which pattern was presented at the input. It is necessary to know the activation of both hidden units.

2. The final state of the network for solving the encoding problem is depicted in Figure 5.7. Can you decipher how the network has solved the problem? For example, why is it that the second output node has a positive bias while all the others have a negative bias?

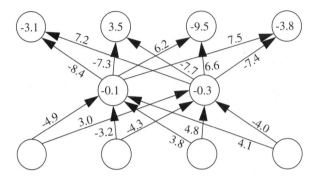

FIGURE 5.7   Network for the solution of the encoding problem in **auto1**.

Exercise 5.4

- At first blush, you might conclude from Exercise 5.3 that the network with 2 hidden units had used up all the distributed representations available—2 units can encode 4 binary patterns—and that more hidden units are necessary to solve the 5 pattern problem. This would be correct if the hidden units were restricted to binary activations. But, of course, they're not! They can take on any *real* value between 0 and 1. This means that there are many possible combinations of hidden unit activations that can be used to represent distinct input patterns (see *Rethinking Innateness*, Chapter 2, Figure 2.17 and associated discussion).

     So the answer is that **auto2** can solve the problem with just 2 hidden units though you might find that it finds a solution quicker with 3. Why do you think this might be?

Exercise 5.5

- With the suggested input pattern (0.6 0 0 0.2), the network does an excellent job of reconstructing a cleaner version of the input. However, as the noise on other input units increases in value the accuracy of completion will deteriorate. When the input is ambiguous, such as when two input units are both fully on or at 0.5, pattern completion will be particularly bad.

Exercise 5.6

*   We say that information in a pattern is redundant if one bit of the input vector can be predicted by the activity of another bit. Of course, in order to conclude that this is (or is not) the case, it is necessary to look at more than one pattern. Hence, redundancy is a property of a set of patterns, not a single pattern.

Exercise 5.7

1. As the patterns are displayed, it may seem that the activity of the first node divides the 8 patterns conveniently into 2 groups. But, of course, this is true of all the nodes in the patterns. Alternatively, you might decide to group the patterns in terms of the number of nodes that are activated, or whether that number is odd or even. Clearly, there are many ways in which the patterns could be categorized.

2. The network should learn to perform the task quite well given that you have used the training parameters suggested in the text.

3. Of course, your estimation of how well the network has done depends on the criteria you set for success!

Exercise 5.8

1. The hidden unit activations for the different input patterns are shown in Figure 5.8, sorted by first, second and third hidden units, respectively. The first hidden unit appears to categorize the patterns neatly into two distinct groups. However, it is unclear from the patterns what the basis of this categorization is. In contrast, the second hidden unit seem to be sensitive to the activity of the first and fourth input units and the third hidden unit to the activity of the second and fifth input units.

2. If you select a subset of the input patterns for the encoding problem, the nature of the task defined for the network is changed—correlations in the activity of different input units across a large set of patterns may not hold for a smaller set of patterns. The smaller set might contain the only

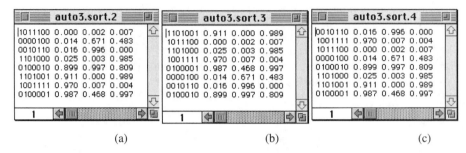

auto3.sort.2				auto3.sort.3				auto3.sort.4		
1011100 0.000 0.002 0.007				1101001 0.911 0.000 0.989				0010110 0.016 0.996 0.000		
0000100 0.014 0.671 0.483				1011100 0.000 0.002 0.007				1001111 0.970 0.007 0.004		
0010110 0.016 0.996 0.000				1101000 0.025 0.003 0.985				1011100 0.000 0.002 0.007		
1101000 0.025 0.003 0.985				1001111 0.970 0.007 0.004				0000100 0.014 0.671 0.483		
0100010 0.899 0.997 0.809				0100001 0.987 0.468 0.997				0100010 0.899 0.997 0.809		
1101001 0.911 0.000 0.989				0000100 0.014 0.671 0.483				1101000 0.025 0.003 0.985		
1001111 0.970 0.007 0.004				0010110 0.016 0.996 0.000				1101001 0.911 0.000 0.989		
0100001 0.987 0.468 0.997				0100010 0.899 0.997 0.809				0100001 0.987 0.468 0.997		

1		1		1	

        (a)   (b)   (c)

FIGURE 5.8   Unit activations sorted on the first (a), second (b) and third (c) hidden units, respectively.

exceptions to the correlation in the larger set. So you shouldn't expect the hidden units to develop identical patterns of activity when trained on a subset of the patterns. Try it!

3.   It is relatively easy to test any hypothesis about the cause of the hidden unit activations—just create some novel data patterns that contain only the target feature you have in mind and see if the hidden units react as you predict (using the **Testing Options...** dialogue box to select the novel data set). Of course, you can always take a close look at the weight matrix in an attempt to evaluate your hypothesis. This approach often works well for small networks, but becomes cumbersome for larger systems.

CHAPTER 6   # Generalization

Why is generalization important?

We rarely train networks on random collections of data. Typically, we
choose the data because they come from some domain in which we are
interested. Sometimes we know in advance what the relationship is
between the input/output pairs and our goal is to see whether the network can discover it. An example of this is the XOR function; we
know that this is a difficult function and we may wish to study the
conditions under which it may or may not be learned. Other times, we
know there is a regular relationship but may not be able to formulate
it precisely; our goal in this case is to use the network as a discovery
device to uncover hidden relationships and make the mapping function explicit. For example, when Lehky & Sejnowski (1990) trained a
network to determine shape from shading information, they were able
to analyze the network afterwards in order to discover the computational primitives used to solve the problem.

   The important point in both cases is that we assume the network
has *extracted some generalization from the training data.* That is the
whole point of training. Whether our purpose is to see whether the
network *can* generalize, or use the network to help discover *what* the
generalization is, we want the network to induce the underlying regularity from the examples given.

How do we know when a network has generalized?

Let us imagine we train a network on some set of data. After some number of training cycles, we observe that the error on the dataset (say, averaged over all the items) has dropped to a low number.

How low is low enough? Well, here we must be careful. Let's say the training set consists of 100 patterns; each pattern is itself a 100-element vector with one of the bits set to 1 and the remaining 99 bits set to 0. That is, the desired targets would look something like:

TABLE 6.1   Sample data

Pattern number	Desired outputs (each vector has 100 elements)
#1	10000...000
#2	01000...000
#3	00100...000
...	...
#99	00000...010
#100	00000...001

Let's also assume that over the course of training, we monitor the RMS error and observe the pattern shown in Figure 6.1. At the outset of training, RMS error is close to 50; after 1,500 training sweeps the error has dropped to approximately 1.0. This seems good; based on the mean error we might feel the network has learned the dataset.

Surprise! In this example, a mean error of 1.0 might actually result from very bad performance. How might this be? Let's think about our training data in detail and ask ourselves what it would take for the network to achieve such a performance. Each output pattern is of the same basic form (100 0's with a different single bit set to 1) so we can pick an arbitrary teacher pattern and focus on that. Let's pick the first pattern: 10000...000. When networks are initially configured, connections are often assigned weights drawn from a uniform distribution ranging say from -1.0 to 1.0. The mean of the weights on incoming lines to a node is thus 0.0. Therefore, net input (the product of weight times activation on sending units) will tend to be close to

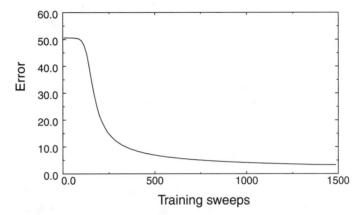

FIGURE 6.1 Hypothetical error plot after training a network to produce the outputs shown above. After 1500 pattern presentations, the error is slightly above 1.0.

0.0. Assuming a sigmoid activation function, nodes will therefore usually have activations close to 0.5 at the outset of learning. (This actually makes sense, from the viewpoint of making it easy for a node to move in either direction.) Given a target pattern of 10000...000, this will yield an initial RMS error of approximately 50.0. (The first unit has an activation of 0.5 when it should be 1.0, giving an error of 0.5; and the remaining 99 units are 0.5 when they should be 0.0, yielding a summed error of 49.5) This is why the error shown in Figure 6.1 starts off close to 50.0.

Now consider the problem facing the network. Each pattern has 99 0's and a single 1, always in a different position. Given the large number of 0's which are present in each pattern, a quick way to reduce error across the entire pattern set would be to turn *all* of the output nodes off. If the network does this, it will get 99 of the output nodes correct (so there will be 0.0 error from them) and only one output node wrong. The total error will thus be 1.0—quite a dramatic improvement over the initial error! This is what we see in the latter portion of Figure 6.1. It should be clear, however, that although the network has in some sense learned an important feature of the training set (i.e., that patterns contain a preponderance of 0's), it hasn't learned the part we might be more interested in (i.e., which bit position has a 1, on any given pattern). The lesson of this example is that we need to think carefully about the criteria for success. If we are going to use a global statistic such as mean RMS error over a pattern set, we must be careful to figure out what performance could give rise to different ranges of error, and what error levels would be indicative of success (however we chose to define it).

Exercise 6.1

- Consider the pattern set described in the previous section (i.e., 100 patterns, each containing 99 0's and a single 1). Let us establish the criterion that to be successful, the network must (a) have output activations not greater than 0.1 on bits which should be 0.0, and (b) have an output activation not less than 0.9 on the bit which should be 1.0. (This is a somewhat stringent criterion; in reality, we might be satisfied with outputs of 0.3 or 0.7.) What error level must we have on a single pattern to be guaranteed that the network's performance meets this criterion?

In previous chapters, you've used a network to approximate a function associating a set of input patterns with a set of output patterns. In the XOR task explored in Chapter 4, for example, a network was trained by presenting all possible input patterns and the appropriate output. In many cases, however, it is preferable to withhold some input/output pairings from the training set, and to use these patterns to test the fully trained network. If the network performs correctly on these novel patterns (i.e., the network has never seen these particular patterns before), then one can feel confident that the network has induced a general function from the specific exemplars on which it was trained, rather than simply "memorizing" the training set. Such a network is said to have *generalized*.

One of the principal factors controlling generalization is the number of hidden nodes available to the network. If a network has sizable internal resources (relative to the task at hand), there will be little pressure on the network to find an efficient, and hence more general, solution. Rather, the network will simply memorize the patterns. (Note that the use of the term "memorize" in this context, while common among connectionist researchers, is somewhat misleading. A network *always* learns *some* function from input to output—in the sense that if you present a novel input pattern to the network, some pattern of activation will be produced across the output units—but the function may not correspond to what the researcher "had in mind.") If a network has too few hidden units, then it will simply be unable to learn the training set. A trade-off is thus established; give the network enough internal resources to learn the training set (to some criterial performance level), but no more (to encourage generalization). Finding this magic number of hidden units usually involves some trial and error.

TABLE 6.2   An incomplete XOR

INPUT		OUTPUT
0	0	0
0	1	1
1	0	1

A second important factor to keep in mind when trying to get a network to generalize is the selection of the training set. The subset of all possible patterns must properly sample the function space or the network won't have any chance of generalizing properly (i.e., to the function you had in mind). For example, what do you think would happen if you tried to get a **2x2x1** network (like the network in

Exercise 6.2

- What function do you think the network would learn from the data in Table 6.2? Would it generalize properly to the missing XOR pattern? Try it. Do you think it's possible to train a network on XOR with less than the full data set?

Figure 4.1) to learn XOR by training it on the three patterns in Table 6.2.

In this chapter, you will:

- Evaluate characteristics of the network's architecture and training data that promote generalization.

- Show how the response properties of the unit's activation function can influence generalization.

- Learn how to perform cluster analyses of hidden unit activations—another technique for getting at the similarity structures discovered by the network.

Analogue addition

Continuous Xor

In Chapter 4 you saw how a three-layer, feedforward network can learn to perform the XOR function via the backpropagation learning algorithm. The XOR task employs both binary inputs and binary outputs. But backpropagation is by no means restricted to binary inputs or outputs. For example, a continuous-input version of the XOR problem can be defined as in Table 6.3. The rule can also be expressed ver-

TABLE 6.3   Continuous Xor

INPUT		OUTPUT
> 0.5	< 0.5	1
< 0.5	> 0.5	1
> 0.5	> 0.5	0
< 0.5	< 0.5	0

bally as follows:

> "Combine pairs of inputs such that if either is greater than 0.5 the output is a 1; however, if both are greater than 0.5 or less than 0.5, then the output is a 0."

Thus, we have created a problem which—instead of having binary inputs and binary outputs—has continuous inputs and binary outputs.

Binary outputs are typical of *classification* problems in which the inputs (consisting either of binary or continuous-valued units) are classified by the network into one of several categories. Problems of

Exercise 6.3

• If we plot the input pairs as points on a Cartesian plane, the space of all possible inputs is the square region bounded by the points (0,0), (0,1), (1,0), and (1,1). The continuous-input XOR partitions this space into two distinct regions. Draw a graph illustrating these regions. Do you think this version of the XOR problem is easier or harder than the binary version for a network to solve? Why?

---

this sort essentially require the network to partition the input space into as many disjoint regions as there are output categories.

In the more general case, input and output units are *both* continuously valued. The role of the network then is to construct a continuous mapping from input values to output values based on the examples provided by the training data. In principle, a three-layer feedforward network is capable of performing *any* desired mapping from input to output. But this ignores the crucial questions of how many hidden units will be required, and how easily and accurately the mapping will be learned. Nevertheless, networks using backpropagation to learn continuous mappings have been quite successful in a number of real-life applications.

The addition problem

In this section, you will examine a very simple example of a network learning to perform a continuous mapping from input to output: analog addition (i.e., the output value is to be the sum of the input values). As with XOR you will have two input nodes, two hidden nodes, and one output node. Since we are simulating an addition process, set the output node to have a linear activation function. To achieve this, just include the line

```
linear = 3
```

in the **SPECIAL:** section of the **.cf** file. This instruction sets all the nodes to the right of the equals sign to be linear units, i.e., their activation is the same as their net input. In this case there is only one output node—node 3. You may think that including linear nodes in the network trivializes the problem. Note however that all input activations must first pass through the layer of nonlinear hidden nodes before reaching the linear output node.

Create a **New Project** and call it **addition**. Let us agree to allow only inputs between 0.0 and 0.5; then we only expect output between 0 and 1. Furthermore, let's restrict the training data to numbers with only a single decimal place. This means that there are 36 possible training patterns. You can test the network on higher precision numbers later. Now you need to create a set of training data. This training set will consist of pairs of numbers between 0.0 and 0.5 and their corresponding sums. Unlike the XOR problem, do not present the network with an exhaustive list of all possible input/output combinations. Instead, devise what you believe is a representative subset of perhaps ten or so possibilities.

Exercise 6.4

- Run the network for 200 epochs[a] with a learning rate of 0.3 and momentum of 0.9. Now test the network. How well has the network learned the training data?

---

a. An epoch consists of a single presentation of all the training examples in the **data** file. Since there are, say, 10 input patterns in **addition.data** then 200 epochs consists of 2000 sweeps. Note that if you choose to**Train randomly** there is no guarantee that all the input patterns will be presented on a given epoch, unless you determine that pattern selection is conducted in a random manner *without replacement* in the **Training options** dialogue box by deselecting the **with replacement** check box.

The deeper question here is how well the network has *generalized* from the training data. After all, the training data only informs the network as to what the input-output mapping should be for a small set of points in the input space. How do you know that the network will give the desired answers for inputs that were not included in the training set?

Another interesting question is how the network actually performs the addition, in view of the inescapable nonlinearity in the two hidden nodes.

Exercise 6.5

- How well does the network add *any* valid pair of inputs? You can create novel pairs of inputs in a new **data** file called **novels.data** and load those patterns into the network using the **Novel data:** option in the **Testing Options...** dialogue box. You do not need to specify an equivalent **novels.teach** file just as long as you switch off the **Calculate error** option in the **Testing Options...** dialogue box. Note however that the test file must end with the extension **.data**. before testing the network. Try a smaller or larger training set; how sensitive is the network to the size of the training set? What if the training set draws most of its examples from cases where both inputs are less than 0.25? How accurately does the resulting network add inputs greater than 0.25?

Exercise 6.6

- Draw the network. Show weights on each of the connections, and put the biases inside the nodes. Can you explain the principle that the network is using to perform the addition? Can you think of an alternative principle, i.e., a fundamentally different approach to getting a network with nonlinear hidden nodes to output a linear combination of its inputs?

Categorization

Next consider a categorization function which takes 4-bit vectors as input and sorts them into two categories. The single output bit is ON when the input vector has exactly two bits ON (two and only two), in all other cases the output bit is OFF. Thus, six of the 16 possible inputs produce a 1 at the output, the rest produce a 0.

Exercise 6.7

---

1. Build a **4x3x1** network. Call the project **gen**. Create the necessary **data** and **teach** files according to the function described above, but withhold two of the 16 patterns as indicated in Table 6.4. Train the network using a learning rate of 0.3 and a momentum of 0.9 for 400 epochs. Test the network. Has the network learned to categorize the patterns correctly? If not, repeat the experiment with a different random seed until it succeeds.

2. Now test the network for generalization using the two patterns which you withheld from the training set. You will need to create a new **data** file and load it into the program with the appropriate set of weights. Has the network generalized properly? Do you think the same thing would have happened if you had withheld *any* two patterns? Test your hypothesis by repeating the experiment with different pairs withheld.

3. Try repeating the experiment with a **4x2x1** network. Does the network learn the training set? It may well not. Try different random seeds until the net converges. Why do you think it's so much more difficult with only two hidden nodes? That is, given that a solution *does* exist with two hidden nodes, why does the network fail so frequently to find it?

An additional hidden unit gives the network another dimension of freedom in traversing the weight space. Thus, the network is less likely to get caught in a local minimum. When you've found a random seed which works with a **4x2x1** network save the weights file. In the next section, you'll explore some techniques for analyzing the network's solution.

TABLE 6.4   Test examples for **gen** project.

INPUT				OUTPUT
0	0	1	0	0
1	0	1	0	1

Network analysis

Once a network has learned a training set to some criterial perform-ance level, the next step is to analyze the solution which the network has found. A variety of techniques are employed to do this, each of which gives valuable clues regarding network behavior. For example, you might probe the network with carefully selected inputs and look at the resultant hidden node and output patterns. This should give you some idea of the manner in which the network is transforming infor-mation. This is analogous to conducting a psycholinguistic experi-ment in which the network is the subject and the probe stimuli are selected to test hypotheses about what the network has learned. You should also examine the network performance across *each pattern* in the training set. Even though the total error may have dropped to cri-terion, the network may in fact be performing very well on some pat-terns, and only moderately well on others.

Exercise 6.8

- Does this differential performance suggest some-thing about the nature of the network's solution?

Once these holistic techniques have been exhausted, however, the next step is to open up the network and look inside. Unfortunately, much like a brain, most networks simply look like mush when you examine their innards. Nonetheless, a couple of feasible alternatives do exist. With a small network, for example, you can simply draw (by hand) a big version of the network, write in the connection strengths and biases, then pass activation through the network by calculating it yourself. This was the technique you used in analyzing the XOR net-work in Chapter 4.

Hand analysis

Try this technique again using the weights file you generated at the end of the section on categorization. Recall that in that experiment a network learned to take a 4-bit input string and (using just two hidden nodes) to output a 1 if the input had exactly two bits turned ON (two and only two). All other inputs produced a 0 on the output.

Exercise 6.9

---

1. Think about this task for a moment. Can you think of a solution to this problem (across two hidden nodes)? That is, under what conditions should each of the two hidden units turn ON, such that the network could properly complete the problem in the next layer of connections (from hidden to output)?

2. Now draw out the network you actually trained and pass several test patterns through the connections. (Remember that you can use the logistic table in Exercise 1.1 on page 8 to calculate the activation from the net input.)

3. Can you characterize the function of each of the two hidden units? Under what conditions do they each turn ON? How does this solution compare to the one you invented?

Cluster analysis

Another technique for analyzing a network solution is to look at how the similarity structure of the input patterns is changed as a result of going from the input to the internal representation. **tlearn** possesses a utility called **Cluster Analysis...** (an option in the **Special** menu) which exists to aid you in this regard. This utility takes as input a set of vectors and performs a hierarchical cluster analysis, drawing a tree diagram of the similarity structure. The **Cluster Analysis...** dialogue box is shown in Figure 6.2. Start by seeing what sort of similarity structure is inherent in your input patterns. Use the **tlearn**

```
┌─────────────────────────────────────────┐
│     Cluster Analysis of Vectors          │
│                                          │
│   Vector file: [                     ]   │
│                                          │
│   Names file: [|                     ]   │
│                                          │
│   ⊠ Display Graphical Cluster Tree       │
│   ☐ Report Clusters and Distances        │
│              Output to                    │
│   ☐ Text            ⊠ Graphics           │
│   ☐ Suppress Scaling                     │
│   ☐ Verbose Output                       │
│                                          │
│  ( Dismiss )   ( Cancel )  ( Execute )   │
└─────────────────────────────────────────┘
```

FIGURE 6.2   **Cluster Analysis...** dialogue box.

editor to generate two files from your **gen.data** file. One file should contain just the 14 four-bit input vectors (call it **gen.inp**) and the other file should contain 14 pattern names. Your files should be identical with those shown in Figure 6.3. To cluster the input vectors open the relevant files in the **Vector file:** and **Names file:** boxes,

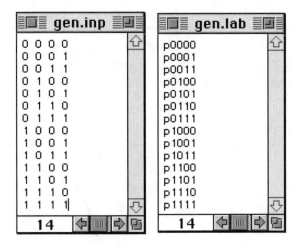

FIGURE 6.3   Sample input files for the **Cluster Analysis...** utility.

select the **Display Graphical Cluster Tree** box and **Execute** the analysis. **tlearn** will display a window containing a cluster analysis on your screen as shown in Figure 6.4.

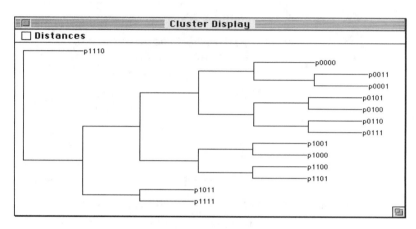

FIGURE 6.4   Cluster analysis of the **gen.data** pattern files.

Exercise 6.10

> • Can you see any similarity structure in the input patterns? Is this structure relevant to the task? Why not?

Now perform a similar analysis for the hidden node activations. You will need to test all the patterns in your training set and save the hidden node activations as you did in Exercise 4.5 on page 85. Then create the necessary files for the cluster analysis as above.

Exercise 6.11

> • Has a meaningful similarity structure emerged at the hidden unit level? Does this agree with what you learned earlier about the function of each of the two hidden nodes?

The symmetry problem

These are the basic techniques of network analysis. Try using them yourself on the following problem, called the symmetry problem. Build a **6x2x1** network, which will take 6-bit inputs and output a 1 if the input pattern is symmetric, i.e., if the last three bits mirror the first three. For example, **0 1 0 0 1 0** and **1 1 0 0 1 1** are symmetric, while **0 1 1 0 1 1** is not. Thus, eight of the 64 possible input patterns are symmetric. Note that it *is* possible for a network to solve this problem with only two hidden units, but it may require some fiddling with the random seed, learning rate, and momentum.

Exercise 6.12

- Once the network has learned the problem, analyze its solution using the techniques discussed in this chapter.

Answers to exercises

Exercise 6.1

- This is a trickier question than it might first seem. The answer is not 10.0, as you might think (assuming that each of 100 bits produces 0.1 error). Other scenarios exist which could give an even lower error but still not satisfy our conditions for success. Suppose, for example, the network produced a 0.0 on all the outputs. This would yield an error of only 1.0—but our condition (b) would not be satisfied. Clearly, only one output activation need be off by more than 0.1. So the highest level of error that meets the desired criterion for a single pattern is just 0.1.

Exercise 6.2

- Unless you are really lucky, the network will learn Boolean OR when trained on the patterns listed in Table 6.2. When you test the network's generalization to the pattern 1 1, it will most likely respond with an output of 1. You are not providing the network with any evidence for it to believe this is not a linearly separable problem (see Exercise 4.6 on page 92). Under what circumstances might you get lucky and have the network generalize to the novel input 1 1 as if it had learned XOR?

Exercise 6.3

- The partitioning of the Cartesian plane for Continuous XOR is shown in Figure 6.5. This is a more difficult problem than Binary XOR because the network has to be more precise about how it partitions the space. Recall from Chapter 4 that the network solved XOR by moving the 4 points around the Cartesian plane so that they could be partitioned appropriately (see Exercise 4.6 on page 92). Now the network must take more points into consideration.

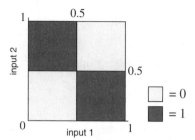

FIGURE 6.5   Partitioning of the Cartesian plane for Continuous XOR.

Exercise 6.4

• Unless you were unlucky and chose a set of training parameters that got the network stuck in a local minimum, you should experience no difficulty in training the network to solve this problem. If the network did get stuck in a local minimum, then just try another random seed to initiate the weight matrix.

Exercise 6.5

• The network does surprisingly well at adding any legal combination of inputs, just so long as you have constructed a training set that is representative of the problem. For example, the network can learn to perform addition for all 36 input patterns when it has only been trained on 5 of them, if you make sure that the training set spans the full range of outcomes possible, i.e., 0.0 through 1.0. However, if you restrict the problem to inputs below 0.25 you will find that the network becomes increasingly inaccurate as the solutions extend above 0.5.

Exercise 6.6

•   When the network is trained for 2000 sweeps with the inputs shown in Figure 6.6, with a learning rate of 0.3, momentum of 0.9 and a random seed of 1 (trained randomly without replacement), the final state of the network is as shown in Figure 6.6.

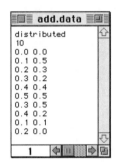
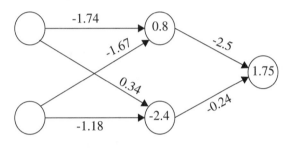

FIGURE 6.6   Training set and weight matrix after 2000 sweeps on the addition problem

The network has discovered a rather clever solution to the problem. Notice that the bias on hidden node 2 is quite large and negative. With the negative connection from the second input node, the second hidden node will remain virtually inactive (output close to 0.0) for any input pattern. The second hidden node, therefore, has no effect on the output activation. In contrast, the first hidden node has a small positive bias and almost identical negative connections from the two input nodes. Consider the case when the input is 0 0. The net input to first hidden node is just the bias, i.e. 0.8. This produces an activity of around 0.7 on this hidden node (see Exercise 1.1 on page 8). When this activation is fed through the negative connection to the output node, it exactly cancels out the positive bias to produce an output of 0.0 (remember the output node is linear). Now consider the input 0.5 0.5 which is the largest legal input to which the network is exposed. In this case, the activity propagating up from the input nodes more or less cancels out the positive bias of the first hidden node to produce a net negative input of -0.87. This produces an activity of around 0.3 on the first hidden node. When this activation is fed through the negative connection to the output node, it counteracts the positive bias to produce an output of 1.0. The first hidden node, the connections feeding into and out of it, and the bias on the output node are doing all the work.

How has the network managed to solve the addition problem given that the hidden nodes are nonlinear? We have just seen that we only need to consider the first hidden node. The minimum net input arriving at this node is -0.87 and the maximum net input is 0.8. Now recall the node's sigmoid activation function shown in Figure 1.3 on page 5. When inputs are restricted to this range the sigmoid function is more or less linear. The network has found a solution which exploits the linear component of the sigmoid curve! In other words, it has effectively turned the nonlinear unit into a linear unit by suitable selection of weights and bias.

Exercise 6.7

- You should be able to crack this one on your own! However, if you experience difficulties finding a configuration of training parameters for the problem with 2 hidden units try a random seed of 6 (without replacement), a learning rate of 0.3 and a momentum of 0.9.

Exercise 6-8

- If the network makes larger errors on some of the input patterns than others, then it probably hasn't learned the appropriate generalization but some other function.

Exercise 6.9

1. It is quite easy to think up a solution to this problem using just 2 hidden units. One hidden unit can have a negative bias that keeps it switched off unless there are two active input units. This hidden unit can be connected to the output with a positive connection. The second hidden unit should have a negative bias that keeps it switched off unless there are 3 or more active input units. However, the connection from this unit to the output is large and negative to counteract the activity propagating from

the first hidden unit. Under these circumstances, the output unit will only fire when exactly 2 input units are active. A simple network like this is shown in Figure 6.7.

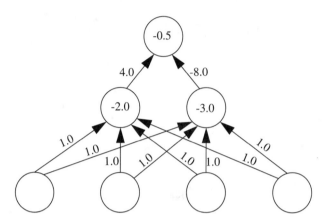

FIGURE 6.7   A hand-wired network for solving the categorization problem.

2.   The solution that your network found by itself might look like that in Figure 6.8. In this example, the first hidden node remains firmly off unless there are at least 3 active input lines. Once this node becomes active, its negative connection to the output node ensures that output is switched off too. The second hidden node has a strong positive bias and remains switched on if there are less than 2 active input lines (note the negative connections between the input nodes and the second hidden node). If the second node is switched on, the strong negative connection to the output node makes sure that the output node is switched off. In other words, the output node is switched off if there are at least 3 or less than 2 active nodes at the input. If there are just 2 active input nodes, the second hidden node switches off but the first hidden node doesn't switch on. In other words, both hidden nodes are dormant allowing the positive bias on the output node to activate it. This solution is quite different to the hand-wired solution in Figure 6.7.

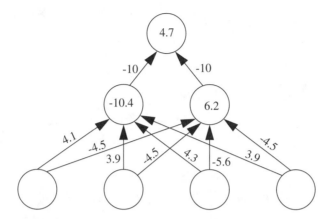

FIGURE 6.8   A self-organized solution to the categorization problem.

Exercise 6.10

• The cluster analysis of the input patterns groups them together as best it can according to their position in the four dimensional input space. So the pattern 1011 is closer to the pattern 1111 than it is to the pattern 0000. This grouping is not relevant to the task that the network is being asked to perform. For example, the network is supposed to group 0011 together with 1100 but these patterns are quite far apart in the four dimensional input space (see Figure 6.4).

Exercise 6.11

• The cluster analysis of the hidden unit activations using the network shown in Figure 6.8 is given in Figure 6.9. Now the network has grouped together all the patterns in which just two input nodes are active (the middle branch of the tree). All patterns with just one input node are placed in the first branch of the tree. Patterns with 3 or 4 input nodes active are placed in the third branch of the tree. This organization corresponds exactly to the activities of the hidden nodes that we analyzed in Exercise 6.9.

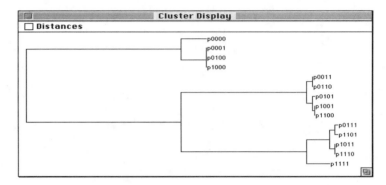

FIGURE 6.9   Cluster analysis of the hidden unit activations in the generalization problem

CHAPTER 7 *Translation invariance*

Introduction

In this chapter you will examine the ability of the network to detect patterns which may be displaced in space. For example, you will attempt to train a network to recognise whenever a sequence of three adjacent 1's occurs in a vector, regardless of what other bits are on or where the three 1's are. Thus, **01110000, 00011100, 10101111** all contain this pattern, whereas **01100000, 10101010, 11011011** do not.

Why might such a problem be of interest? This task is schematic of a larger set of problems which we encounter constantly in everyday life and which are sometimes referred to as examples of *translation invariance*. When we recognize the letter **A** on a page, or identify other common objects regardless of their spatial location, we have solved the problem of perceiving something which has undergone a spatial translation. (We can also usually perceive objects which have been transformed in other ways, such as scaling, but here we address only displacement in space.) We carry out such recognition without apparent effort, and it probably does not even occur to us that a pattern which has been moved in space ought to be particularly difficult to recognize.

In fact, dealing with spatial translations is quite difficult for many visual recognition schemes, and we will find that it is also hard for networks. You will look at one solution, but first you will demonstrate that the problem is hard and try to understand just what the basis of the difficulty is.

In this chapter, you will learn:

- How to configure a neural network so that its hidden nodes have constrained receptive fields (instead of receiving connections from all the units in the previous layer).
- Show that receptive fields and unit groupings are important for solving the problem of translation invariance in a neural network.

Defining the problem

Create a **New Project** called **shift**. Build an **8x6x1** network. The training set should contain the following patterns (remember that the vector elements in **shift.data** should have spaces between them:

```
11100110 11101001 01110101 01110000
10111011 00111010 11011100 01011101
01101110 10001110 10010111 00100111

10011010 11001101 10000010 01001100
01000000 01101000 00100011 10100100
00110110 10010110 10011000 00010000
00011011 00010001 01010000 01011011
00000100 00001101 00000011 00001001
```

Be sure you have entered these patterns exactly. Check your files!

Of the 32 patterns, the first 12 contain the target string **111**, while the last 20 do not. Thus, your **shift.teach** file will have 1 as output for the first 12 patterns and 0 for the last 20. Try running the simulation with a learning rate of 0.3 and a momentum of 0.9 for 200 epochs (this means 6400 sweeps). Be sure to choose the **Train Randomly** option. Feel free to experiment with these parameters. Test the network on the training data.

Exercise 7.1

- Has the network learned the training data? If not, try training for another 200 epochs or run the simulation with a different random seed.

Now test the network's ability to generalize. Create a new **data** file containing novel patterns (call it **novshift.data**) using the following eight input patterns, four of which contain the target string and four of which do not.

```
00000111  11100100  11101100  01110011
10110001  10001101  11011011  01101101
```

Test the network's response to these novel patterns:

Exercise 7.2

---

1. How well has the network generalized? Use the clustering procedure you learned in Chapter 6 on the hidden node activation patterns of the training (not test) data.

2. Can you tell from the grouping pattern something about the generalization which the network has inferred?

Relative and absolute position

In the previous simulation you saw that although it is possible to train the network to correctly classify the training stimuli, the network does not generalise in the way you want. Note that this does not mean that the network failed to find some generalisation in the training data, simply that the generalisation was not the one you wanted.

It is important to try to understand why spatial translation is such a difficult problem. The basic fact to be explained is that although bit patterns such as **111000000** and **000111000** look very similar to us, the network sees them as very different. The absolute bit pattern (e.g., whether the first three bits are **0** or **1**) is a more important determinant of similarity than the relative bit pattern (e.g., whether any three adjacent bits are **0** or **1**).

One way to think about why this might be so is to realize that these bit patterns are also vectors, and that they have geometric interpretations. A 9-bit pattern is a vector which picks out a point in 9-

dimensional space; each element in the vector is responsible for locating the point with regard to its own specific dimension. Furthermore, the dimensions are not interchangeable.

Consider a concrete example. Suppose we have a 4-element vector. To make things a bit easier to imagine, we will establish the convention that each position in the vector stands for a different compass point. Thus, the first position stands for North, the second for East, the third for South, and the fourth for West. We will also let the values in the vector be any positive integer. We might now think of the following vector: **1 2 1 1** as instructions to walk North 1 block, then East 2 blocks, South 1 block, and West 1 block. Consider the 4x4 city block map shown in Figure 7.1 and start in the lower left corner and see where you end up.

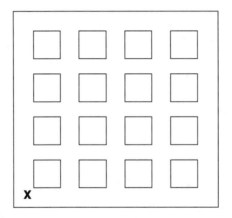

FIGURE 7.1   A geometric interpretation of input vectors

Now let us rotate the numbers in this vector, so that we have something that looks similar (to our eye): **2 1 1 1**. However,—and this is important—we keep the convention that the first number refers to Northern movement, etc. Now we have a different set of instructions. Start again in the lower left corner and see where you end up this time. Not surprisingly, it's a different location. You are probably not surprised because it is obvious that going 2 blocks North and then 1 block East is different from going 1 block North then 2 blocks East.

The situation with the network is similar. Each position in a vector is associated with a different dimension in the vector space; the dimensions are analogous to the compass points. The vector as a whole picks out a unique point.

When the network learns some task, it attempts to group the points which are picked up by the input vectors in some reasonable way which allows it to do the task. (Thinking back to the city-block metaphor, imagine a network making generalizations of the form "all points in the north-west quadrant" versus "points in the south-east quadrant.") Thus, the geometric interpretation of the input vectors is not just a useful fiction; it helps us understand how the network actually solves its problems, which is by spatial grouping.

You can now see why shifts or translations in input vector patterns are so disruptive. You can shift the numbers around, but you cannot shift the interpretation (dimension) that is associated with each absolute position. As a result, the shift yields a vector which "looks" very different to the network (i.e., picks out a different point in space) even though to the human eye the two patterns might seem similar.

These shifts do not necessarily make it impossible for the network to learn a task. After all, in the previous simulation the network succeeded in classifying various instances of shifted **...111...** patterns. XOR is another problem which is successfully learned by the network even though the vectors which have to be grouped are as different as possible (think about the locations of the points picked out in a square by **00**, **11**, **01**, **10**, and the groupings that are necessary for the task). The network can overcome the dissimilarity. (The spatial contortions necessary to do this, however, usually require a hidden layer; the hidden layer takes on the job of reorganizing the spatial structure of the input patterns into a form which facilitates the task.) The point is that the classification solution is not likely to generalize to novel patterns, just because what seemed like the obvious basis for similarity to us (three adjacent **1**'s) was for the network a dissimilarity which had to be ignored.

Again, it is worth thinking about why this problem might worry us. Many human behaviors—particularly those which involve visual perception—involve the perception of patterns which are defined in *relative* terms, and in which the *absolute* location of a pattern in space is irrelevant. Since it is the absolute location of pattern elements which is so salient to the networks you have studied so far, you now want to see if there are any network architectures which do not have this problem. In the next simulation, you will study an architecture which builds in a sensitivity to the relative form of patterns. (This architecture was first described by Rumelhart, Hinton & Williams, (Chapter 8, PDP Vol. 1) and used in the task of discriminating T from C, regardless of the orientation and position of these characters in a visual array. You may wish to review that section of the chapter before proceeding.)

Receptive fields

Your guiding principle in creating an architecture to solve the shift invariance problem will be this: *Build as much useful structure into the network as possible*. In other words, the more tailored the network is to the problem at hand, the more successful the network is apt to be in devising a solution.

First, you know that the pre-defined target string is exactly three units in length. Therefore, design your network so that each hidden node has a *receptive field* spanning exactly three adjacent input nodes. Hidden nodes will have overlapped receptive fields, staggered by one input. This means that if **111** is present at all in the input, one of the 6 hidden units will receive this as its exclusive input, and the two neighboring hidden units on each side will receive part of the pattern as their input (the hidden unit immediately to the left sees two of the **1**'s, and the unit beyond that sees only one).

Now what about the issue of shift invariance? Each receptive field has a hidden unit serving it exclusively. (For more complicated problems, we might wish to have several hidden units processing input from a receptive field, but for the current problem, a single unit is sufficient.) Let us designate the receptive field weights feeding into a hidden unit as RFW1, RFW2, and RFW3 (Receptive Field Weight 1 connects to the left-most input, RFW2 to the center input, and RFW3 to the right-most input). We have 6 hidden units, each of which has its own RFW1, RFW2, and RFW3. We can require that the RFW1's for all 6 hidden units have *identical* values, that all RFW2's be identical, and all RFW3's be identical. Similarly, the biases for the 6 hidden units will also be constrained to be identical. Thus, **11100000** will activate the hidden node assigned to the first receptive field in exactly the same way that **01110000** will activate the hidden node assigned to the second receptive field. Finally, since all 6 hidden units are functionally identical, we want the weights from each hidden unit to the output unit to be identical. Figure 7.2 shows the architecture we have just described.

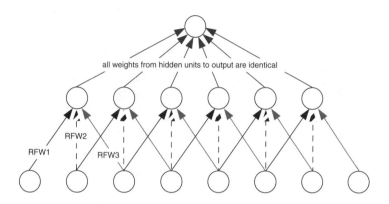

FIGURE 7.2   Network architecture for the translation invariance problem

How can we ensure that this occurs? Our solution will be to initialize each receptive field hidden unit to have an identical set of weights (compared with other units), and then to average together the weight changes computed (according to the backpropagation learning algorithm) for each connection and to adjust each such weight only by the average change for that position. Fortunately, the **tlearn** simulator has an option that will perform the necessary averaging automatically, but it is still necessary to tell the program which weights are to be made identical. To do this, we need to employ the **groups** option in the **.cf** file. All connections in the same group are constrained to be of identical strength.

The **NODES:**, **CONNECTIONS:** and **SPECIAL:** entries for the **.cf** file that you will need for this exercise are shown in Figure 7.3. Notice how each connection or set of connections is identified as belonging to one of 5 groups. The changes made to one weight will be the average of the changes computed by backpropagation for all the

```
NODES:                CONNECTIONS:              SPECIAL:
nodes = 7             groups = 5                selected = 1-6
inputs = 8            1-6 from 0 = group 1      weight_limit = 0.1
outputs = 1           7 from 0
output node is 7      7 from 1-6 = group 2
                      1 from i1 = group 3
                      1 from i2 = group 4
                      1 from i3 = group 5
                      2 from i2 = group 3
                      2 from i3 = group 4
                      2 from i4 = group 5
                      3 from i3 = group 3
                      3 from i4 = group 4
                      3 from i5 = group 5
                      4 from i4 = group 3
                      4 from i5 = group 4
                      4 from i6 = group 5
                      5 from i5 = group 3
                      5 from i6 = group 4
                      5 from i7 = group 5
                      6 from i6 = group 3
                      6 from i7 = group 4
                      6 from i8 = group 5
```

FIGURE 7.3   `shift2.cf` file for translation invariance problem.

weights with which it is grouped. (We have formatted this file in three columns; in the real file, the material shown in the **CONNECTIONS**: column would immediately follow the **NODES**: section, etc.)

Exercise 7.3

1. Draw a diagram of the **8x6x1** network, and indicate those weights and biases which are constrained to be identical. Check this with the way that **tlearn** has configured the network.

2. Train the network for 2000 epochs (64,000 sweeps) with a learning rate of 0.3 and momentum of 0.9. (Use random pattern selection.) Has the network learned the training set? If not, try training the network with a different random seed.

The fact that you have successfully trained this new network on the training data does not necessarily imply that the network has learned the translation invariance problem. After all, we saw that the **8x6x1** network in the first part of this chapter (the **shift** project) also learned the training data; crucially, its failure to generalize in the way we wanted was what told us that it had not extracted the desired regularity. (It's worth pointing out again that the network has undoubtedly generalized to *some* function of the input, but simply not to the one we wished.) Therefore you must test this network with new data.

Exercise 7.4

> • When you have successfully trained the network, test its ability to generalize to the novel test patterns. Has the network generalized as desired?

It is possible that on the first attempt, your network may not have generalized correctly (but this is not common); if it fails, retrain with a different starting seed.

Finally, it is worth looking at the network to try to understand its solution. This involves examining the actual weights and drawing the network, with weight values shown, in order to determine what the network's solution is. When you do this, work backwards from the output unit: Ask yourself under what conditions the output unit will be activated (indicating that the target pattern of ...**111**... was found). Take into account both the output unit's bias and the activation received from the 6 hidden units. Then ask what input patterns will cause the hidden units to be activated, and what input patterns will cause them to turn off.

Exercise 7.5

> • Examine the contents of the weight file. Draw out the weights for one hidden node, the weight connecting it to the output unit, and the biases for the hidden unit and output unit. (These should be identical across different hidden units). Do you understand the network's solution?

Answers to exercises

Exercise 7.1

• It may take a few attempts, but generally this network will succeed in learning the training data after a few attempts. If the network has learned the correctly, the first 12 outputs will be close to 1.0 (but values as low as 0.70 may be acceptable) and the last 20 outputs will be close to 0.0 (again, actual outputs will only approximate 0.0).

Exercise 7.2

1. The first four patterns in **novshift.data** all contain the ...**111**... pattern, whereas the last four do not. If the network has generalized as desired, then the first four outputs will be close to 1.0 and the final four will be close to 0.0. This is not likely to be the case. (It is barely possible that your network, by chance, stumbles on the solution you want. If so, if you run the network another four or five times with different random seeds, you are not likely to replicate this initial success.)

   To do the clustering on the training data hidden unit patterns, you will need to go back to the **Network** menu, and in the **Testing Options...** submenu and for the **Testing set**, select **Training set [shift.data]**. Then, again in the **Network** menu, choose **Probe selected nodes**. This will run the network once more, sending the hidden unit outputs to the **Output** display window. Delete any extraneous material you have in the **Output** display and in the **File** menu, use **Save As...** to save the hidden unit activations in a new file called **shift.hidden**. Before clustering, you will also need to prepare a labels file (called **shift.lab**) which is identical to **shift.data**, but with the first two (non-pattern) lines removed, and with all spaces deleted. (Since this can be cumbersome, we have already prepared a file with this name and placed in the folder for Chapter 7.) If you now run the **Cluster Analysis** (found in the **Special** menu; send output to graphics), you might see something that looks like Figure 7.4:

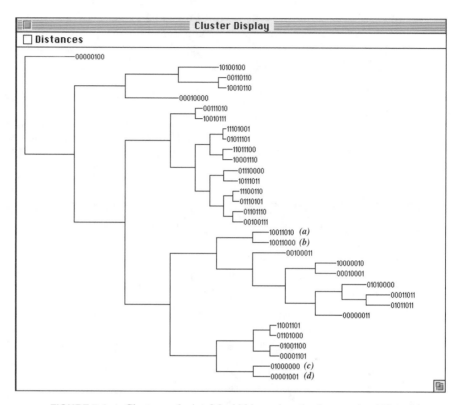

FIGURE 7.4   Cluster analysis of the hidden unit activations on the shift problem

2. Notice that all the patterns which contain ...**111**... are clustered together on the same branch; this tells us that the hidden unit patterns produced by these inputs are more similar to each other than to any other inputs. That is what allows the network to treat them the same (i.e., output a **1** when they are input). However, if you look closely, you may also see that the principle by which inputs are grouped appears to have more to do with the degree to which patterns share **1**'s and **0**'s in the same position. This is particularly apparent for patterns *(a)* and *(b)*, and patterns *(c)* and *(d)*. The fact that inputs which do not contain the ...**111**... target pattern also happen to have many **0**'s in their initial portions—and that it is this latter feature which the network is picking up on—

should lead us to predict (correctly) that the network would classify the novel test pattern **00000111** on the basis of the initial **0**'s, and ignore the fact that it contains the target.

Exercise 7.3

1. After having drawn your network, display the architecture using the **Network Architecture** option in the **Displays** menu. You will see a diagram which looks like that shown in Figure 7.5.

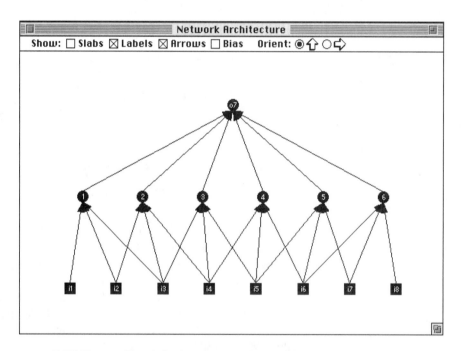

FIGURE 7.5   Translation invariance network architecture

2. We train the network for 2000 epochs simply to ensure that the weights have converged on relatively stable values. This will produce cleaner outputs and make subsequent analysis of the network a bit easier.

Exercise 7.4

- The network should have generalized successfully so that it recognizes the first four patterns in **novshift.data** as containing the target (i.e., the network output is close to 1.0), and the last as not containing the target (i.e., the output is close to 0.0). If this is not the case, retrain the network using a different random seed, or experiment with different values for the learning rate and momentum.

    You may find it useful to keep the **Error Display** active while you are training. If you see that the error does not appear to be declining after a while, you may choose to abort the current training run prematurely and restart with different values. After a while, you may begin to develop a sense of what error plots will ultimately lead to success and which ones are destined to result in failure.

Exercise 7.5

- Figure 7.6 shows the receptive field weights for one hidden unit. (All other input-to-hidden and hidden-to-output weights should be the same.) The biases are shown within each unit.

    Working backwards, we note that the output unit has a strong positive bias. By default, then, it will be on (signalling detection of the target pattern). So we then have to ask the question, What will turn the output unit off? We see that the input which the output unit receives from the 6 hidden units is always inhibitory (due to the weight of -8). However, the output unit's bias is sufficiently large (43) that *all* of the hidden units must be activated in order for their combined effect to be great enough to turn off the output (since -8x5 generates only -40). But if we look at the hidden units' biases, we see that they are strongly positive (14). This means that by default the hidden units *will* be activated. The hidden units' default function is therefore to suppress firing of the output. Overall, the default case is that the output says there is no target present.

    What will cause the output unit to fire, then? If a single hidden unit is turned off, then the remaining hidden units' output will not be sufficient to turn off the output unit and it will fire, indicating detection of the target. So what can turn off a hidden unit? Since the hidden unit bias is 14, and each input weight has an inhibitory weight of -5, all three inputs must be present to turn off a hidden unit, which then releases the output

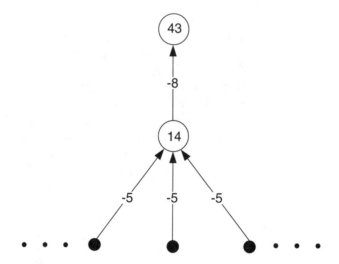

FIGURE 7.6   Receptive field weights for a hidden unit in the translation invariance
network

unit from suppression and turns it on. If two or fewer adjacent inputs are
present, they will be insufficient to turn off the hidden unit.

This may seem complicated at first, but it actually is a very sensible
solution!

## CHAPTER 8

Simple recurrent networks

Introduction

In Chapter 7, you trained a network to detect patterns which were displaced in space. Your solution involved a hand-crafted network with constrained weights (so that different hidden nodes could benefit from the learning of others). Now turn your attention to the problem of detecting patterns displaced in *time*.

This problem requires a network architecture with dynamic properties. In this chapter, you'll follow the approach of Elman (1990) which involves the use of *recurrent* connections in order to provide the network with a dynamic memory. Specifically, the hidden node activation pattern at one time step will be fed back to the hidden nodes at the next time step (along with the new input pattern). The internal representations will, thus, reflect task demands in the context of prior internal states. An example of a recurrent network is depicted in Figure 8.1.

In this chapter, you will:

- Train a recurrent network to predict the next letter in a sequence of letters.

- Test how the network generalizes to novel sequences.

- Analyze the network's method of solving the prediction task by examining the error patterns and hidden unit activations.

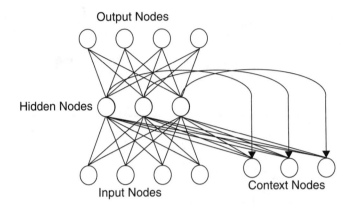

FIGURE 8.1   A simple recurrent network.

File configuration

Starting on the Training Data

Imagine a letter sequence consisting of only three unique strings:

ba dii guuu

For example, part of the sequence might look like this:

babaguuudiiiguuubadiiidiibaguuuguuu...

Note that the sequence is only semi-random: the consonants occur randomly, but the identity and number of the following vowels is regular (i.e., whenever a **d** occurs, it is always followed by exactly two **i**'s). If we trained a dynamic network to predict successive letters in the sequence, the best we could expect would be for the network to say that all three consonants are equally likely to occur in word-initial position; but once a consonant is received, the identity and number of the following vowels should be predicted with certainty.

A file called **letters** exists in your **tlearn** folder. It contains a random sequence of 1000 words, each 'word' consisting of one of the

3 consonant/vowel combinations depicted above. **Open...** the **letters** file. Each letter occupies its own line. Translate these letters into a distributed representation suitable for presenting to a network. Create a file called **codes** which contains these lines:

```
b 1 1 0 0
d 1 0 1 0
g 1 0 0 1
a 0 1 0 0
i 0 0 1 0
u 0 0 0 1
```

Now with the **letters** file open and active, select the **Translate...** option from the **Edit** menu. The **Translate** dialogue box will appear as shown in Figure 8.2. Set the **Pattern file:** box to **codes** and

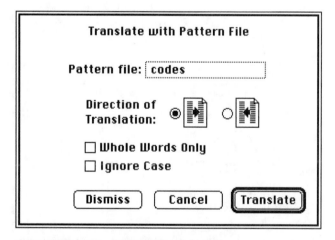

FIGURE 8.2   The **translate** dialogue box.

check that the **Direction of Translation** is from left to right. Then click on **Translate**. Your **letters** file is translated into rows and columns of binary digits. (Note that both the **letters** and **codes** files must be in the same directory or folder.) Each row consists of one of the sequences of 4 digits taken from the **codes** file. **Translate** has replaced every letter in the **letters** file with a pattern vector. Every occurrence of a letter in the **letters** file which corresponds with a letter in the first column of the **codes** file is replaced by the sequence of alphanumeric characters to the right of the first column in the **codes** file[1]. **tlearn** asks you for the name of file in which to save the translated data. To avoid overwriting the existing **letters** file, call it **srn.data**. Next copy this file to a file called **srn.teach** and edit the file, moving the first line to the end of the file. The **srn.teach** file is now one step ahead of the **srn.data** file in the sequence. Complete the **teach** and **data** files by including the appropriate header information.

Now build a **4x10x4** network with 10 context nodes. These special context nodes will store a copy of the hidden node activation pattern at one time step and feed it back to the hidden nodes at the subsequent time step (along with the new input). The **srn.cf** file is shown in Figure 8.3. Notice that the context nodes are specified as nodes **15-24** and are set as **linear** in the **SPECIAL:** section.

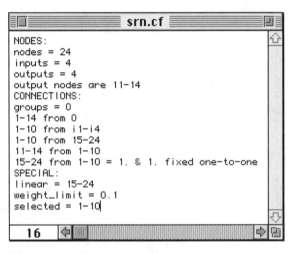

```
srn.cf

NODES:
nodes = 24
inputs = 4
outputs = 4
output nodes are 11-14
CONNECTIONS:
groups = 0
1-14 from 0
1-10 from i1-i4
1-10 from 15-24
11-14 from 1-10
15-24 from 1-10 = 1. & 1. fixed one-to-one
SPECIAL:
linear = 15-24
weight_limit = 0.1
selected = 1-10

16
```

FIGURE 8.3   The **srn.cf** file.

Nodes **1-10** receive connections from the context nodes. However, the context nodes also receive connections from nodes **1-10** (last line

---

1. Notice that you could translate your transformed file back again if you wish by using the alternative **Direction of Translation** in the **Translate...** dialogue box.

of the **CONNECTIONS:** section). This line also indicates that these 'copy-back' connections have a minimum & maximum value of 1.0, that they are fixed and that they are in a one-to-one relation to each other, i.e., node 1 sends a connection only to node 15, node 2 only sends a connection to node 16, etc.

Exercise 8.1

---

1. Why are the the hidden nodes and the context nodes fully connected in one direction but not in the other? Why do you think the context nodes are set to be linear?

2. Draw a diagram of your network. You may wish to use slabs to indicate layers (rather than drawing individual nodes). Indicate those weights which are fixed at 1.0.

You are now in a position to train the network. However, before you do so you may as well create a set of patterns for testing after you have trained the network.

Exercise 8.2

---

1. What patterns would you include in your test set? Construct the test set and call it **predtest.data**.

2. Set the learning rate parameter to 0.1 and momentum to 0.3. Train the network for 70,000 sweeps (since there are 2993 patterns in **srn.data**, this is approximately 23 epochs), using the **Train sequentially** option in the **Training Options** dialogue box. It is imperative that you train sequentially and not train randomly. Why?

3. To see how the network is progressing, keep track of the RMS error. Why do you think the RMS error is so large?

Exercise 8.2

---

4. Test the network using the **predtest.data** file. How well has the network learned to predict the next element in the sequence? Given a consonant, does it get the vowel identity and number correctly?

5. What does it predict when a consonant is the next element in the stream?

Run through these test patterns again but plot a graph of the error as the network processes the test file. To do this you will need to construct a **predtest.teach** file and make sure that the **Calculate error** box is checked in the **Testing options...** dialogue box. You should be able to see how the error declines as more of a word is presented. Thus, error should be initially low (as we predict an **a** following the first **b**, then increases when the **a** itself is input and the network attempts to predict the beginning of a new sequence.

Furthermore, if you look at the bit codes that were used to represent the consonants and vowels, you will see that the first bit encodes the C/V distinction; the last three encode the identity of the letter. When individual output bits are interpretable in this way, you might wish to look not only at the overall error for a given pattern (the sumsquared error across all bits in the pattern) but at the errors made on specific bits.

Exercise 8.3

---

1. What do you notice about the network's ability to predict the occurrence of a vowel versus a consonant as opposed to specific vowels and consonants?

2. Finally, investigate the network's solution by examining the hidden node activation patterns associated with each input pattern. Perform a cluster analysis on the test patterns. Can you infer something from the cluster analysis regarding the network's solution?

Answers to exercises

Exercise 8.1

1. The downward connections from the hidden units to the context units are not like the normal connections we have encountered so far. The purpose of the context units is to preserve a copy of the hidden units' activations at a given time step so that those activations can be made available to the hidden units on the next time step. So what we need, therefore, is a mechanism which will serve to copy, on a one-to-one basis, each of the hidden unit's activations into its corresponding context unit. We do this by setting up downward connections which are one-to-one and fixed (i.e., not changeable through learning) at a value of 1.0. That ensures that the input to each context unit is simply the activation value of its "twin" hidden unit. In addition, we define the context units themselves to have linear activation functions. This means that their activation is simply whatever is input to them, without undergoing the squashing which occurs with units which have (the normal) logistic activation function. The result is that the context unit preserves the value exactly.

2. Your network should look like the one shown in Figure 8.4. We have used slabs to indicate banks of units to avoid clutter. "Distributed" means that all units from one layer are connected to all other units in the second layer; "one-to-one" means that the first unit in one layer is connected only to the first unit in the second layer, etc.

Exercise 8.2

1. What we wish to verify is that the network has learned (a) that each consonant is itself unpredictable, and so should predict all three equally likely, whenever a consonant is expected; and (b) that each consonant predicts a specific vowel will occur a certain number of times. To do this, we need a test set which contains one occurrence of each possible 'word', e.g., the sequence **b a d i i g u u u**. As with the training data, you will have to be sure to use the **Translate...** option

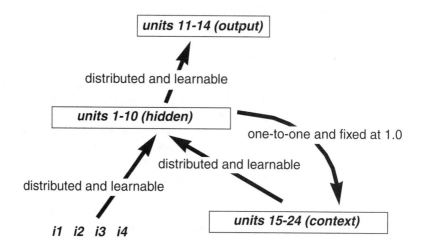

FIGURE 8.4   Schematic of network for the letter prediction task

from the **Edit** menu to convert the letters to vector form. Or, since there are such a small number of patterns involved, you could create the vectors by hand (just be sure not to make mistakes!).

2. The whole point of this chapter is that there are *sequential dependencies* in the data. Obviously, then, if the network is to learn these sequential dependencies, it must experience the data in their correct sequence. Random pattern presentation would mean that a **d** might sometimes be followed by an **i**, other times by an **a**, or **u**, or even another consonant. **Train sequentially** ensures that the network sees the patterns in their correct order, as they appear in the .**data** file.

3. In Figure 8.5 we show the error after 70,000 sweeps of training. From the fact that the error has not declined very much and asymptotes at a relatively high level, we might think that the network has not learned much. And in many cases, this would be a reasonable conclusion.

    But there is another possibility to consider before we give up in despair. Remember that the training data are only partially predictable. In fact, there are two things which the network can learn. (1) Once it sees a consonant, the identity and number of occurrences of the vowel that follows can be predicted with certainty; (2) a corollary of knowing

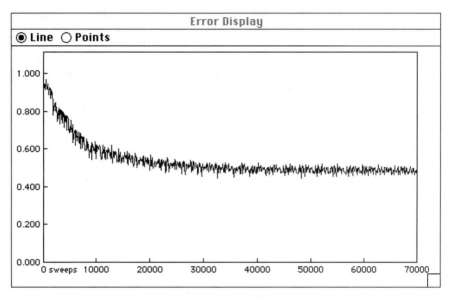

FIGURE 8.5   RMS error for the letter prediction task over 70,000 sweeps

the number of vowels which follows is that the network can know when to expect a consonant. However, the identity of the consonants themselves has been chosen at random. So—unless the network memorizes the training set (which is not likely)—we have given the network a task which inevitably will continue to generate error. Since the error plot averages the error over 100 sweeps and shows a data point only once every 100 sweeps, we are actually combining the error when vowels occur (which should be low) with the error when consonants occur (which should be high). Overall, the average declines as the network learns to get the vowels right; but since it can never predict the consonants, the error can never go to 0.0.

4. When you test the network, the output will be shown in vector form. To interpret this output, you will have to consult the **codes** file (shown also on page 153) and work backwards[2]: Find the letter which most closely resembles the vector output from the network. Remember that the network's output is a prediction of what the next letter will be, so the very first line should be similar to the second letter in the **predtest.data** file, which is an **a**, coded as **0 1 0 0**.

5. After the last vowel in a sequence has been input, the network prediction should be wrong—it may look something like one of the consonants, but the precise identity should not be correct.

Exercise 8.3

1. We can define Vowel as any vector with a **0** in the first position, and Consonant as any vector with a **1** in the first position (just because that is the way we set up the vectors in the **codes** file to begin with). So now let us look to see what the network is predicting, concentrating only on the first bit position. In one of our own runs, this is what we got (the vowel which should be predicted is shown in parentheses to the left):

(a)	0.000	0.969	0.083	0.004
(d)	0.977	0.345	0.409	0.220
(i)	0.000	0.019	0.985	0.005
(i)	0.006	0.031	0.919	0.039
(g)	0.987	0.444	0.224	0.331
(u)	0.000	0.013	0.010	0.984
(u)	0.001	0.008	0.039	0.979
(u)	0.138	0.099	0.065	0.853
(b)	0.995	0.417	0.512	0.132

Even though the network isn't able to predict *which* consonant to expect, it does clearly know when to expect *some* consonant, as evidenced by a high activation on the first bit.

2. To do a cluster analysis of the hidden unit activations produced by the inputs in **predtest.data**, we need first to clear the **Output** display (if it is open) (in the **Edit** menu, **Select All**), and then in the **Network** menu, **Probe selected nodes**. This will place the hidden unit activations in the **Output** display. With that window active, go into the **Edit** menu and **Save As...** a file called

---

2. There is also an **Output Translation...** utility available in the **Special** menu and described in Appendix B (page 263 and page 289) which can be used to read output vectors as letters. An example of the use of the **Output Translation...** utility is described in Chapter 11 on page 212.

predtest.hid. This will be the vector file we cluster. We will also need a **Names** file, which we can create by hand and call **predtest.lab.** This file should consist of the inputs which produced the hidden unit activations, i.e., the letters **b a d i i g u u u**, one per line. Since there are several instances of some of the vowels, we might wish to mark them individually, e.g., **b a d i1 i2 g u1 u2 u3** so that we can tell them apart on the plot. Then choose **Cluster Analysis** from the **Special** menu, checking **Output to Graphics**. When we did this, this is the plot we got:

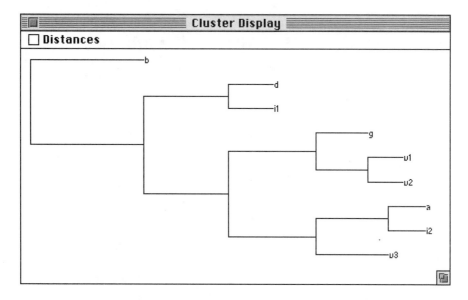

FIGURE 8.6   Cluster analysis of the hidden unit activations for letters in the prediction task

At first, some aspects of this plot may appear a bit odd. For instance, why should the hidden unit patterns when a **g** is input be similar to **u**? And why should the hidden unit patterns in response to **a**, **i2**, and **u3** be similar to each other (as shown in the bottom branch)?

One thing to remember is that the network's task is *prediction*. We might expect it, therefore, to develop similar hidden unit patterns when it has to predict similar outputs. Viewed this way, in terms of what each of the inputs above *predicts*, the clustering makes much more sense. The

a, **i2**, and **u3** in the bottom branch all have something very important in common: They are the final vowels in their subsequence. When the network encounters any **a**, it knows that a consonant should follow. Similarly, when the network encounters the second **i** or third **u** in a sequence, the next letter will be a consonant. In the second branch from the bottom, the **g**, **u1**, and **u2**, all predict a **u**. And the **b** at the top is different from all other letters because it alone predicts a following **a**. So the network's internal representations reflect what it has learned about the classes of inputs with regard to what they predict.

## CHAPTER 9

Critical points in learning

Introduction

An intriguing phenomenon in the developmental literature is the observation that many children undergo what appear to be two distinct phases in generalization. In the first stage, they are able to generalize to novel instances, but the generalizations are in some sense restricted to the types of data they have encountered. In the second stage, the generalization ability is unbounded.

For instance, Karmiloff-Smith has noted that children will often undergo a phase where they can demonstrate partial mastery of the principle of commutativity. They know that 2+5 is the same as 5+2, and are able to extend this to other sums which are similar in the sense of involving small numbers (e.g., 7+3, 8+4, etc.). They will fail to recognize, however that 432+85 is the same as 85+432. Thus they are able to generalize a basic notion of commutativity, but the generalizations are limited to sums which may be novel, but which must closely resemble the sums they have encountered before.

At a later stage, of course, children (and adults) recognize that for any arbitrary pair of numbers, their sum is the same regardless of their serial order. At this stage—"true generalization"—the principle of commutativity seems to exist independently of the examples through which it was learned. This is a level of abstraction which seems to be qualitatively different from the knowledge possessed at the earlier stage.

It is not apparent at first glance that a neural network might exhibit behavior which is similar. Neural networks frequently are able to learn functions, and these can then be applied to novel data. *A pri-*

ori, however, one might expect the level of generalization to be a graded and continuous phenomenon which increases smoothly with experience. One would not necessarily expect the sort of two-stage behavior which is observed in children. This is an important issue, because networks use learning devices which operate continuously in time. That is, the learning mechanism does not change. In children, on the other hand, the appearance of dramatic changes in behavior is often taken as evidence for a qualitative change in the underlying learning mechanism (e.g., through reorganization, reanalysis, redescription, etc.).

Surprisingly, such behavior *can* be observed in networks. Networks can undergo qualitative changes in their performance of just the sort which are demonstrated by children who are learning commutativity.

In the following simulation you will explore the generalization ability of a network over the time course of its learning a function. Rather than teaching the network commutativity, you will teach it to tell the difference between 'odd' and 'even' (the parity function, extended over time). You will find that the network shows two kinds of generalization which closely resemble the phases observed in children. Furthermore, you will localize the "magic" point in time where the first type of generalization gives way to the second.

The task

In this task, you will present the network with short sequences of 1's and 0's. Each bit is presented one at a time; the network's task is—at each point in time—to tell you whether an odd or even number of 1's has been input. Sequences will randomly vary in length between 2 and 6. After each sequence is done, the network will be reset (the activations of the context nodes set to 0) and you will start with a new sequence. For example (**/** marks the end of a sequence and beginning of the next):

```
input:   1 1 0 1 / 0 0 / 0 1 0 0 1 0 / 1 1 1 ...
output:  1 0 0 1   0 0   0 1 1 1 0 0   1 0 1 ...
```

In the **tlearn** folder there are three files called **teach**, **data** and **reset**. **data** contains the input sequence consisting of 1000 randomly generated sequences; **teach** contains the teacher signal for each input; and **reset** contains the list of numbers identifying sequence beginnings (in the example above, the file would have **0, 4, 6, 12,** etc.). When **tlearn** gets to the beginning of a new sequence, it resets all the context units to 0.0, so that the prior state is gone. Edit these files to create the files needed to run a simulation with a recurrent network containing one input node, 3 hidden nodes and 1 output node.

Exercise 9.1

• How many context nodes will the network contain?

Call the project **train1**. You will also need to rename the **reset** file to **train1.reset**. This file will need to be edited so that it contains a header line—an integer specifying the number of time stamps to follow. Each time stamp is an integer specifying the time step at which the context nodes are to be completely reset. The time stamps should appear in ascending order.

Training the network

You will go through several passes of training the network. First, train for about 10 epochs (which means 24880 sweeps). Use a **weight_limit** of 0.1 and select the **Use X-entropy; Log RMS** option in the **Training Options...** dialogue box. Set the learning rate and momentum parameters to quite low levels (say, 0.05 and 0.2 respectively). After you have trained the network, you will have a new file, called **train1.24880.wts**.

To test the network, create a data file which contains a single sequence of 100 1's. This is a convenient pattern because:

• It is by far longer than any sequence the network has encountered; you can look to see at what point the network starts to fail.

• It will be easy for you to see visually how the network is doing.

What is cross-entropy, and why use it?

The cross-entropy measure has been used as an alternative to squared error. Cross-entropy can be used as an error measure when a network's output nodes can be thought of as representing independent hypotheses (e.g., each node stands for a different concept), and the node activations can be understood as representing the probability (or confidence) that each hypothesis might be true. In that case, the output vector represents a probability distribution, and our error measure—cross-entropy—indicates the distance between what the network believes this distribution should be, and what the teacher says it should be.

There is a practical reason to use cross-entropy as well. It may be more useful in problems in which the targets are 0 and 1 (though the outputs obviously may assume values in between.) Cross-entropy tends to allow errors to change weights even when nodes saturate (which means their derivatives are asymptotically close to 0).

The correct answer will be an alternating sequence of 1's and 0's (1 0 1 0 1 0), just because the sequence flip/flops back and forth between beyond "odd" and "even."

However, although we can see this output in the **Output** window, **tlearn** does not currently have the capability of displaying the output graphically; only the error can be graphed (in the **Error Display** window). We can use the following trick to generate an error display which actually shows what the output looks like. Create a teacher file for the test sequence which consists of 100 1's. This teacher file is not actually the "correct" one (which would be an alternating sequence of 1's and 0's); however, if the network is actually producing the right output (the alternating 1010.... sequence), then the error that will be displaying with this teacher file will itself be an alternating sequence of 0101....[1]

If the network has not learned to generalize at least beyond the longest training sequence (i.e., 6 inputs), repeat the training starting from scratch, but using a different random seed.

If all has gone well, you will find that a network trained for 10 epochs will generalize its odd/even knowledge to sequences longer than 6, but not much longer. Eventually it gets lost and can't keep

---

1. For example, if the network is performing correctly, its first output is 1; this is compared with the teacher file's first pattern, which is 1, giving an error of 0. If the network's second output is 0, that is compared with the teacher file's second pattern, which is also 1, so the "error" is 1-0=1. Thus, what is displayed is not really the error, but the network's output subtracted from 1, which should—if the network has learned the task—be an alternating sequence of 0101....

Exercise 9.2

---

1. Create suitable data and teacher files for your test set, as described in the text. Do you need a **reset** file for the test sequence?

2. Test the network, plotting the error in the **Error Display**. Use the trick described in the text (see Footnote 1). How well has the network learned to keep track of odd/even? For how many input bits is it successful?

3. Has it gone beyond the length of the longest training sequence?

4. Is there some point beyond which it fails?

5. How would you characterize the network's generalization performance?

---

track of whether the longer sequence is odd or even. Note that the performance gradually decays: the output activation is more clearly right (close to 1.0 or 0.0) at the beginning, but gradually attenuates. Nonetheless, there is some point at which it just doesn't know.

(It is possible that you may find networks which either fail completely to learn at all, or succeed completely on the 100-item test string. This reflects the ways in which different learning parameters may interact with each other, as well as the initial random weights generated by the seed you use. Catching the network at just the point where it has partially learned may therefore take some time. If you wish, you may use the weights file we provide in your **tlearn** folder; this is called **train1.demo.24880.wts**. You can use this file for testing purposes by selecting it as the weights file to load with the **Testing Options...** menu. You can also use these weights as a starting point for continued learning by loading them within the **Training Options...** menu.)

Now repeat the training procedure from scratch. This time, train for 15 epochs (37320 sweeps). (Again, if you have not gotten good results with your own networks, use the one provided in **train1.demo.24880.wts**.)

After 15 epochs of training the network should have learned to make the odd/even distinction for the entire 100-bit test sequence. It

Exercise 9.3

---

1. How well has the network learned to keep track of odd/even?

2. Is there some point beyond which it fails?

3. Has it learned to make the odd/even distinction for the entire 100-bit test sequence?

4. How would you characterize the network's generalization performance?

---

does so with great confidence: The activation of the output node should look like a saw-tooth wave, swinging sharply between 1 to 0.

An interesting question now is how does the network go from the first stage to the second? Is there a gradual increase in ability? Or is the change abrupt?

In order to discover this, you would have to gradually hone in on the point where the network change occurs. We know already that generalization is incomplete at 10 epochs. We know that it appears to be perfect at 15 epochs. However, to be sure, we should really test the network with a much longer test sequence (e.g., a sequence of 1,000 consecutive 1's). You would then have to (painstakingly) locate the number of epochs where things change by training for different numbers of epochs, carefully picking your tests to locate the change point. Thus, do a series of experiments in which you train for a number of epochs, run your test, and examine the end of the test output sequence. If the output is flat, you know the network has not learned to generalize to 1,000 bits. If the output alternates and achieves stable values, then you might surmise the network has learned to generalize to at least 1,000 bits (and probably, it turns out, forever).

Exercise 9.4

- Whether or not you actually run the extended test in order to localize the "turning point," what do you think happens at the point where the network shifts from not being able to perform the task for any length string, to the the state where it can? The performance seems to exhibit a qualitatitive change. Is this matched by an underlying change which is better described as being qualitative or quantitative in nature?

Answers to exercises

Exercise 9.1

- Since context units are usually used to hold copies of hidden unit activations, there should be as many context units as hidden units; in this case, that means 3.

Exercise 9.2

1. The point of the test sequence is to test the ability of the network to generalize to a very long string—in this case, a string of 100 1's and 0's. Since the test pattern is one unbroken sequence, we do not need any resets.

2. After training for 10 epochs, test the network with the **test1.data** provided in the **tlearn** folder for Chapter 9, with the **Error Display** window visible. You should see something like that shown in Figure 9.1. Remember that we are using the trick (see Footnote 1) of subtracting network output from 1. The correct response would be an alternating sequence of 010.... which, when graphed, would look like a sawtooth wave. We can count any value less than 0.5 as in the right direction when the output should be 1.0 (and, similarly, as correct if the answer should be 0.0 and the output is greater than 0.5). In our simulation, the network's output has this shape for the first 10 inputs. Your simulation result might vary somewhat, but we would expect performance which is roughly comparable to this.

    If you fail to find a network which performs in this way, you may choose to use the weights file we provide in the **tlearn** folder, called **train1.demo.24880.wts**.

3. Since 10 inputs (speaking now of the first 10 which the network gets right) is a longer sequence then the network has been trained on, the network appears to have generalized beyond the training data.

4. The generalization is limited however. After 10 inputs the network loses track of whether the sequence is odd or even; it goes "brain-dead" on the 11th input.

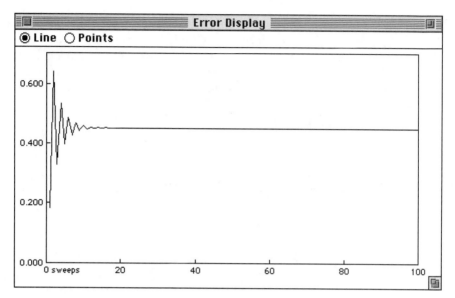

FIGURE 9.1   Error curve for the temporal parity problem when tested after 10 training epochs

5.   The fact that the network succeeds in dealing with sequences which are longer than those it has been trained on suggests that it has succeeded in generalizing the odd/even function. There are two interesting respects in which this generalization is limited, however. First, the generalization is partial; the network gives approximately correct responses for longer strings provided the length is not too great. Past some point (in our example, the 11th input), the network fails. Second, if we interpret the magnitude of the output as an indicator of the network's "confidence," then it seems that as the test string increases in length, beyond that seen in training, the network's confidence decreases steadily. Thus, the generalization is not only partial, but graded, diminishing as the input resembles the training examples less and less.

Exercise 9.3

1. With 15 epochs of training the network should correctly respond to all 100 inputs. (If this is not true for you, load in the **train1.demo.24880.wts** file we provide, and train for an additional 5 epochs—i.e., another 12,440 sweeps) using our weights file as the starting point.)

2. The network should succeed for the entire 100 input test string. The error we get is shown in Figure 9.2.

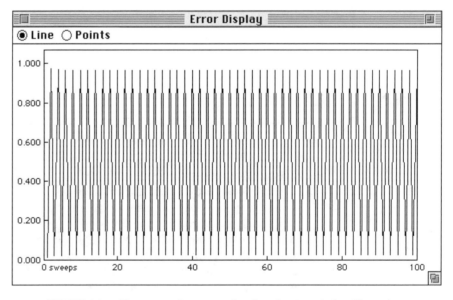

FIGURE 9.2   Error curve for temporal parity when tested after 15 epochs

3. Yes, the network correctly keeps track of the odd/even distinction for the entire test sequence.

4. This network has also learned to generalize, but the generalization appears qualitatively different than when it only has 10 epochs of training. First, the network generalizes well beyond the length of the training data. Second, the network's output does not degrade with increasing length. It appears "confident" of its response throughout the entire sequence. Given this lack of degradation, we might reasonably infer that

the network's generalization is probably absolute: It ought to perform perfectly for strings of indefinite length. (If we really wanted to be sure, however, we ought to test the network at least on strings of length 1,000 or even 100,000.).

Exercise 9.4

*   The key to understanding what is happening lies in recognizing two things:

    The first fact is that the network's memory (which is what it's relying on to know whether the current state is "odd" or "even") depends on (1) the current activations of the hidden units, which must somehow encode the current state; and (2) the recurrent connections (context-to-hidden weights), which allow the network to retain information over time (consider a worst case scenario where these connections are 0; then nothing from the past would be remembered). But notice that the current activations of the hidden units themselves are changed by the recurrent weights, which serve as multipliers on the inputs to the hidden units. So the recurrent weights are a critical factor in learning this task.

    Second, remember that the hidden units' activations are a nonlinear function of their inputs. That means that within certain ranges of magnitude, inputs which vary by a great deal may produce hidden unit activations which differ by very little. Within other ranges, however, slight differences in the magnitude of inputs may produce large differences in activation.

    Taken together, these two facts are the beginning to understanding what happens when the network transitions from its limited generalization to absolute generalization. The problem is simply that until weights are learned which are of a magnitude (and the right sign) to ensure that when hidden unit activations are fed back, their values are of a sufficient magnitude to be retained over time.

    We can illustrate this with an example drawn from Chapter 4 of the companion to this handbook, *Rethinking Innateness*. We will use a simpler network to make the issue clearer; you should be able to extrapolate from this example to what is happening with the odd/even network.

    Let us imagine that we have a network with one hidden unit, one input, and one output, as shown in Figure 9.3.

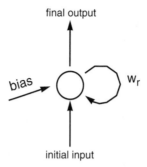

final output

bias w_r

initial input

FIGURE 9.3   A one-node network which receives an initial input; the input is then removed, and processing consists of allowing the network to fold its activation back on itself through the recurrent weight. After some number of iterations, we examine the output.

Let us imagine what would happen if the recurrent weight has a value of w_r= 1.0 and there is a constant bias of b= –0.5 Then if we start with the node having an initial activation of 1.0, on the next cycle the activation will be given by Equation 9.1,

$$a(t+1) = \frac{1}{1 + \exp(-(a(t) - 0.5))} = \frac{1}{1 + \exp^{-0.5}} = 0.62 \qquad \textbf{(EQ 9.1)}$$

or 0.62. If this diminished value is then fed back a second time, the next activation will be 0.53. After 10 iterations, the value is 0.50—and it remains at that level forever. This is the mid-range of the node's activation. It would appear that the network has rapidly lost the initial information that a 1.0 was presented.

This behavior, in which a dynamical system settles into a resting state from which it cannot be moved (absent additional external input) is called a *fixed point*. In this example, we find a fixed point in the middle of the node's activation range. What happens if we change parameters in this one-node network? Does the fixed point go away? Do we have other fixed points?

Let's give the same network a recurrent weight w_r = 10.0 and a bias b = –5.0. Beginning again with an initial activation of 1.0, we find that now the activation stays close to 1.0, no matter how long we iterate. This makes sense, because we have much larger recurrent weight and so the input to the node is multiplied by a large enough number to counteract the damping of the sigmoidal activation function. This network has a fixed point at 1.0. Interestingly, if we begin with an initial activation of 0.0, we see that also is a fixed point. So too is an initial value of 0.5. If we start with initial node activations at any of these three values, the network will retain those values forever.

What happens if we begin with activations at other values? As we see in Figure 9.4, starting with an initial value of 0.6 results over the

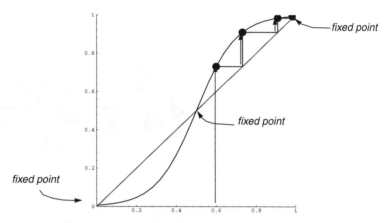

FIGURE 9.4   If a recurrent unit's initial activation value is set to 0.6, after successive iterations the activation will saturate close to 1.0. An initial value of 0.5 will remain constant; an initial value of less than 0.5 will tend to 0.0 (assumes a bias of -5.0 and recurrent weight of 10.0).

next successive iterations in an increase in activation (it looks as if the node is "climbing" to its maximum activation value of 1.0). If we had started with a value of 0.4, we would have found successive decreases in activation until the node reached its fixed point close 0.0. Configured in this way, our simple one-node network has three stable fixed points which act as basins of attraction. No matter where the node begins in activation space, it will eventually converge on one of these three activation values.

The critical parameter in this scheme is the recurrent weight (actually, the bias plays a role as well, although we shall not pursue that here). Weights which are too small will fail to preserve a desired value. Weights which are too large might cause the network to move too quickly toward a fixed point. What are good weights?

Working with a network similar to the one shown in Figure 9.3, we can systematically explore the effects of different recurrent weights. We will look to see what happens when a network begins with different initial activation states and is allowed to iterate for 21 cycles, and across a range of different recurrent weights. (This time we'll use negative weights to produce oscillation; but the principle is the same.) Figure 9.5 shows the result of our experiment.

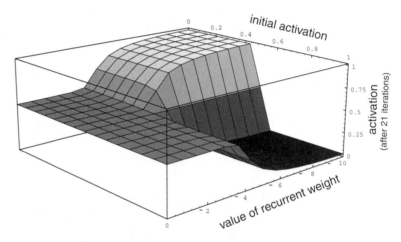

FIGURE 9.5   The surface shows the final activation of the node from the network shown in Figure 9.3 after 21 iterations. Final activations vary, depending on the initial activation (graphed along the width of the plot) and the value of the recurrent weight (graphed along the length of the plot). For weights smaller than approximately -5.0, the final activation is 0.5, regardless of what the initial activation is. For weights greater than -5.0, the final activation is close to 0.0 when the initial activation is above the node's mid-range (0.5); when the initial activation is below 0.5, the final activation is close to 1.0.

Along the base of the plot we have a range of possible recurrent weights, from 0.0 to -10.0. Across the width of the plot we have different initial activations, ranging from 0.0 to 1.0. And along the vertical axis, we plot the final activation after 21 iterations.

This figure shows us that when we have small recurrent weights (below about -5.0), no matter what the initial activation is (along the width of the plot), we end up in the middle of the vertical axis with a resting activation of 0.5. With very large values of weights, however, when our initial activation is greater than 0.5 (the portion of the surface closer to the front), after 21 iterations the final value is 0.0 (because the weight is negative and we iterate an odd number of times, the result is to be switched off; with 22 iterations we'd be back on again). If the initial activation is less than 0.5, after 21 iterations we've reached the 1.0 fixed point state. The important thing to note in the plot, however, is that the transition from the state of affairs where we have a weight which is too small to preserve information to the state where we hold on (and in fact amplify the initial starting activation) is relatively steep. Indeed, there is a very precise weight value which delineates these two regimes. The abruptness is the effect of taking the nonlinear activation function and folding it back on itself many times through network recurrence. This phenomenon (termed a bifurcation) occurs frequently in nonlinear dynamical systems.

To return to the example in this chapter, at the point in training where the network is able to generalize the odd/even solution forever, it has undergone a bifurcation in its weight values. The underlying change involves a process which operates continuously and without sharp transitions; but the effect on behavior is dramatic and appears abrupt.

CHAPTER 10

Modeling stages in cognitive development

The balance beam problem

In this chapter, you will replicate one of the earliest attempts to use neural networks to model cognitive development. McClelland (1989) used a feedforward network to simulate children's performance on the balance beam task. This task was first introduced by Inhelder & Piaget (1958) as a test of children's understanding of reversible operations. More recently, the balance beam task has been studied extensively by Siegler (1981). In this task, children are shown a balance beam with varying weights on either side and at varying distances from the fulcrum (see Figure 10.1). They are asked to judge which side will go

FIGURE 10.1   The balance beam task

down when the beam is released or if it will balance. Siegler has argued that children pass through a series of stages in which their responses appear to be determined by a succession of rules which make differential reference to the dimensions of weight and distance. Initially, children focus exclusively on the heaviest object in making a decision: They choose the side with the heaviest weight, irrespective of its distance from the fulcrum. If the weights are the same children judge the beam to balance. In the second stage of development children start to pay attention to distance, but only under those conditions where the weights are equal on either side of the fulcrum. They judge that the weight which is further from the fulcrum will go down. In the third stage, they can make correct judgements when either of the dimensions are equal on each side of the fulcrum. So if the weights are equal they will judge the one further from the fulcrum to go down. If the distances are equal they will judge the heaviest weight to go down. If one side has more weight and the other more distance, their responses are confused. Finally, some individuals discover (mostly through explicit teaching) the concept of *torque*. Siegler (1981) demonstrated that a large proportion (approximately 85%) of children's responses to the balance beam task fitted into one of these 4 categories.

McClelland's model also exhibits stage-like behavior. However, it achieves this behavioral profile without being told what the rules are for the different stages of development. Furthermore, the changes in the network that give rise to stage-like behavior are continuous. In other words, no new mechanisms are invoked to explain the onset of a new stage. This demonstration is at odds with the standard Piagetian view that distinct stages of development are governed by qualitatively different mental operations (see *Rethinking Innateness*, Chapter 3 for a more detailed discussion).

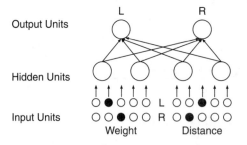

FIGURE 10.2 The balance beam network

Setting up the model

Network architecture

The neural network that McClelland used is illustrated in Figure 10.2. The input to the network is divided into two channels—a channel that represents the weights of the two objects on either side of the fulcrum and a channel that represents the distance of two objects from the fulcrum. There are 5 possible weight values for each object and 5 possible distance positions for each object. Weight and distance values are represented for each object by a 5-place input vector in a localist fashion, i.e. only a single input unit is activated for any weight or distance value.

Weight and distance values are arbitrarily assigned to single input units. The network is not told explicitly which units correspond to large weights or distances nor which units represent the weight or position of the left or the right object. The network must discover these implicit structures for itself. Each bank of input units projects to separate banks of two hidden units. The 4 hidden units project to two

Exercise 10.1

- Why do you think it is important that the network is divided into distinct channels, as opposed to permitting full connectivity from all input units to all hidden units?

output units. The activities of the output units are used to interpret the performance of the network in evaluating which side of the beam goes down for any particular combination of weights and distances. If the activity of one output unit exceeds the activity of the other unit by a stipulated amount, then the 'side' whose unit is most active 'goes down'. Otherwise, the beam balances.

Exercise 10.2

- Create a .cf to define a network architecture of this type. Call it bb.cf.

The training environment

There are 625 possible combinations of weights and distances for the model of the balance beam used in this simulation. Training proceeds through the successive presentation of individual balance beam problems. The network's output is compared to the correct output for that problem and the discrepancy between the actual output and the desired output is used to generate an error signal which is used by backpropagation to adjust the connection strengths in both channels of the network. The network is tested at regular intervals on a range of weight/distance combinations.

Exercise 10.3

1. What do you think an input vector in the `.data` looks like? What does the corresponding `.teach` output vector look like?

2. What output vector will you use for those cases in which the beam balances?

You will train the network on all 625 weight-distance combinations. To save you time, the input and output vectors for the balance beam task are already set up for you in your **tlearn** folder. The files are called **bb.data** and **bb.teach**. Assuming you have created an appropriate **bb.cf** file, start a **New Project** and call it **bb**. **tlearn** will automatically open all the files you need to run your model.

Take some time to look at the training files and examine whether your answers to Exercise 10.3 are correct. Notice that both files contain more than 625 training vectors. If you look at the end of the files you will see that a number of the patterns are repeated several times. These are just those patterns in which the distance values on either side of the fulcrum are identical. In fact, the training files contain 5 repetitions of all balance beam problems in which the distance is the same on either side of the fulcrum and just one example of every other balance beam problem. Later in this chapter, we will consider why this modification of the training environment is necessary.

Testing data

You will evaluate the performance of the network at regular intervals by testing it on a representative subset of the patterns on which it was trained. The test data (found in **bb_test.data**) contain 24 problems subdivided into groups corresponding to the stages of development discovered by Siegler.

1. 4 problems represent balanced beams, i.e., the weights and distances on either side of the fulcrum are identical.

2. 4 problems in which the distance from either side of the fulcrum is the same but the weight varies.

3. 4 problems in which the weight on either side of the fulcrum is the same but the distance varies.

4. Weight and/or distance are in conflict, e.g., the weight may be greater on one side but the distance is greater on the other:

    i.   4 problems in which the side with the greater weight goes down— *conflict weight*.

    ii. 4 problems in which the side with the greater distance goes down— *conflict distance*.

    iii. 4 problems in which the beam balances, i.e., the torques are the same on both sides.

Exercise 10.4

1. Examine the **bb_test.data** file and make sure you're clear how each pattern is classified within this taxonomy of balance beam problems. You will need to use this classification to evaluate performance later.

2. In the meantime, determine which problems the network should get right in each of the four stages outlined by Siegler.

Training the network

Now train the network randomly for 10,000 sweeps using a learning rate of 0.075 and a momentum of 0.9. Set the **weight_limit** parameter to 1.0 in the **bb.cf** file. Use batch learning by setting the **Update weights every:** training option to 100 sweeps. Dump the weights every 400 sweeps (in the **Training Options** dialogue box, check the **Dump weights every** box and set the value of **Sweeps** to 400). You may find it useful to observe the change in RMS error as you train the network, so open the **Error Display**.

Testing the network

As a result of training the network, you will have created a series of 20 weight files which you can use to test network performance at different stages of learning. These weight files will be called **bb.401.wts**, **bb.801.wts**, **bb.1201.wts**, ..., **bb.10000.wts**. Use the **bb_test** files to evaluate the network's capacity to make balance beam judgements.

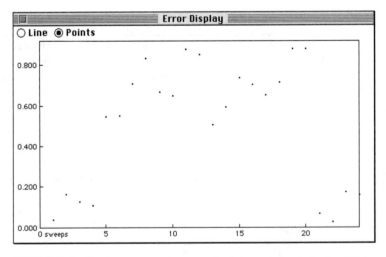

FIGURE 10.3   RMS error on test patterns for balance beam task after 401 sweeps

In the **Testing Options** dialogue box, set the weights file to **bb.401.wts** and the testing set file to **bb_test.data**. You will probably want to save your output in a file so check the **Append output to file:** box and specify a file name. Now **Verify the network has learned**. **tlearn** should output 24 pairs of numbers indicating the output activations for the 24 representative balance beam problems defined by **bb_test.data** If you have checked the **Calculate error** box in the **Testing Options** dialogue box, you can also determine the RMS error for the individual balance beam problems. The **Error Display** should resemble that in Figure 10.3.

Exercise 10.5

• Notice that the RMS error is lowest for the first 4 patterns and last 4 patterns in the test set. Why do you think this is the case? *Hint: Refer to the listing of the test patterns in the section on "Testing data".*

Now examine your output data. You will probably discover that most of the output activations are in the range 0.5 ± 0.2. Almost certainly, none of the output units that should have an activity of 1.0 or 0.0 have achieved their target values.

Exercise 10.6

• Why do you think the output units are in their mid-range of activity?

What constitutes successful performance in this network? We know that it is unrealistic to expect output activations to reach their absolute values of 1.0 or 0.0 (see Exercise 3.3 on page 63 if you want to refresh your memory about this issue), so how are we to decide which side of the beam goes down in those cases when it doesn't balance. The solution that McClelland suggested to this problem is to look at the *difference in activation between the two output units*. He proposed that if one output unit exceeds the activation of the other output unit by more than 0.33, then the 'side' with the highest activation is judged to go down.

Exercise 10.7

---

1. Do you think there is anything special about a difference of 0.33 ?

2. Evaluate the output activities from the network after 401 sweeps of training. How many of the test problems has the network got right when judged according to the current difference criterion?

3. What stage of development do you judge the network to be in at this stage of training. Would your evaluation change if you changed the difference criterion?

4. Repeat your evaluation of network performance with the other weight files (**bb.1201.wts**, **bb.1601.wts**, etc.). Can you discern a sequence of stages that the network passes through in attempting to solve this problem?

Explaining network performance

Consider again the network architecture depicted in Figure 10.2. The network's response to a given balance beam problem is evaluated by comparing the activities of the output units. If the left-hand output unit is criterially more active than the right-hand output unit then the network is interpreted as predicting that the left-hand side of the beam will go down—and *vice versa*. The activity of the output units is determined by the activity on the hidden units and the strength of the connections from the hidden units to the output units. Assume that the left-hand hidden unit in the weight channel receives strong and positive connections from the input units representing the value of the weight on the left-hand side of the beam. Likewise, assume that the right-hand hidden unit receives strong and positive connections from the input units representing the value of the weight on the right-hand side of the beam.

The pattern of connections between the input units and the hidden units forces a complementary pattern of connections from the hidden units to the two output units. In particular, the left-hand hidden unit will grow strong excitatory connections to the left-hand output unit and strong inhibitory connections to the right-hand output unit. The right-hand hidden unit will form connections to the output units which are opposite to those of the left-hand hidden unit. Consequently, when the weights on either side of the beam are equal the activity feeding into the output units from the hidden units will cancel each other out. An analogous distribution of connections in the distance channel of the network will permit the hidden units to influence the output units in a similar fashion.

Exercise 10.8

- Examine the weights file for the trained network. Does the organisation of the connections resemble this description?

The stage-like behavior of the model is directly attributable to the network discovering solutions to each of these problems at different points in its training. Recall that the training set contains repetitions of balance beam problems where the distance dimension is identical on both sides of the fulcrum. Under these conditions of training, the network is more likely to use information about the weights on either side of the fulcrum to decide which side goes down than if all types of balance beam problem are represented equally often. The greater frequency of variation in weight values offers the network more opportunities to extract predictive correlations between weight values and balance beam solutions. In other words, the network can better reduce the error on the output units for a greater number of input patterns by strengthening the connections between the input units, hidden units and output units on the weight channel. Changes are also made to the connections on the distance side of the network. However, these changes are smaller since the distance dimension varies less frequently and appears to be less valuable to the network in predicting the correct output.

Changes to the connections during the early stages of training are slow because the initial random state of the network's connections inhibits the discovery of the relation between weight value and correct output. Some learning occurs, but is slowed down in much the same fashion as learning on the distance channel is slowed down by

Exercise 10.9

> •   Examine the weights files at different stages in train-
>   ing. Do the connections in the 'weight' channel
>   grow faster than the connections in the 'distance'
>   channel?

---

the lower frequency of variation in distance values. However, even small changes in the connections on the weight channel improve the network's efficiency in exploiting the high degree of correlation between weight value variation and correct output, resulting eventually in a sudden change in the magnitude of the strength of connections in the weight channel. Once the strength of the connections in the weight channel is adequate to produce the criterial activity differences on the two output units, the network can be considered to have represented the weight dimension. At this point in learning, the network has mastered the first category of response identified by Siegler (1981), i.e., the weight values exclusively determine the response of the network.

During the next phase of training, the strength of the connections on the distance channel are too weak to influence decisively the relative magnitude of activity on the output units, even when the weight values are identical. During this period, the performance of the network remains stable in relation to the type of solution it offers to the balance beam problem. That is to say, the network's performance remains characteristic of Siegler's first stage. Nevertheless, throughout this period of training, the distance connections continue to be gradually strengthened. The transition to the second stage occurs when the distance weights are strong enough to produce criterial activity differences on the output units when the weight values are equal. The network now pays attention to distance under those conditions when the dominant dimension of weight is equal on both sides of the balance fulcrum (Rule 2).

Again, network performance remains stable with respect to this category of response just so long as the strength of the distance connections do not approach the strength of the weight connections. However, with further training the distance connections become stronger such that the distance dimension influences network performance under conditions of unequal weight values (Rule 3). In some cases, the strength of the connections in the two channels may be sufficiently coordinated so that the network can solve balance beam problems in which weight and distance are in conflict with each other (Rule 4).

Exercise 10.10

1. Do you think it is reasonable to manipulate the training files so that weight values turn out to be a better predictor of network performance than distance values? Can you think of a justification for this manipulation?

2. How important is it that the weight and distance dimension are separated at the input to the hidden unit level? Repeat the simulation with a fully connected network and report your results.

Answers to exercises

Exercise 10.1

- In fact, it is *not* critical that the network is divided into two channels. However, it makes it a lot easier for the network to find a solution to the problem because the hidden units are essentially being told ahead of time what type of job they are supposed to carry out. When the inputs representing *weight* values project only to two hidden units then these hidden units will come to represent primarily information about weight—likewise for the *distance* units. Keeping information about weight and distance separate at the hidden unit level is equivalent to guaranteeing that the network treats these dimensions as orthogonal while trying to decipher a solution to the problem. Hence, the internal representation of a weight is preserved across the hidden units irrespective of its distance from the fulcrum. In this fashion, *weight* and *distance* are only allowed to interact at the output.

  When you have worked through this chapter you might like to try training the network with full connectivity from the input units to the hidden units and compare your results to the network in which the weight and distance channels are segregated.

Exercise 10.2

- The **bb.cf** file should contain the following information:

  ```
  NODES:
  nodes = 6
  inputs = 20
  outputs = 2
  output nodes are 5-6
  CONNECTIONS:
  groups = 0
  1-2 from i1-i10
  3-4 from i11-i20
  5-6 from 1-4
  ```

```
1-6 from 0
SPECIAL:
selected = 1-6
weight_limit = 1.00
```

Exercise 10.3

1. The network contains 20 input units consisting of four 5-place vectors
    that represent the 5 values of distance and weight on either side of the
    fulcrum. A typical input vector might be:

    ```
    0 1 0 0 0 1 0 0 0 0 1 0 0 0 0 1 0 0 0 0
    ```

    Let's suppose that the first 5 units represent the value of the weight
    on the left side of the fulcrum and units 6 to 10 represent the value of the
    weight on the right side of the fulcrum. Similarly, units 11 to 15 repre-
    sent the distance of the left weight from the fulcrum and units 16 to 20
    represent the distance of the right weight from the fulcrum. We use a
    localist representation where a single unit indicates the weight or dis-
    tance value. Furthermore, we decide (arbitrarily) that within each set of
    5 units, weights (or distance) increase from left to right. You can use
    many other schemes but this is the one adopted in the sample file
    bb.data. Notice also that the input vector depicted above could also
    be represented in the localist fashion:

    ```
    2,6,11,16
    ```

    In this case the balance beam will go down on the left (the left
    weight is heavier) so we need to give the network an unambiguous
    teacher signal that will indicate that the left output unit should be far
    more active than the right output unit. We will assume in all cases where
    the beam does not balance that the activation of the unit representing the
    side that goes down has a target of 1 and the other side an activation of 0.
    In the example above, the teacher signal will be:

    ```
    1 0
    ```

2. When the beam balances we want the activations of the output units to be identical. Which level of activation should we choose for the teacher signal in this situation? Remember that most combinations of weight and distance on one side of the fulcrum can result in an up or a down response, i.e., a target activation of either 1 or 0 for the appropriate output unit. An intermediate level of activity for a balanced beam would be 0.5 for both output units giving the teacher signal:

`0.5 0.5`

Notice that this output activity is achieved if the net activity entering each output unit is zero (see Figure 1.3 on page 5).

Exercise 10.4

1. You might find it useful to draw the balance beam with weights and distances marked off for each of the problems listed in the **bb_test.data** file. Table 10.1 shows examples of the different types of problems in the **bb_test.data** file.

TABLE 10.1 Different types of problems in the balance beam task and the percentage of correct responses to these problems when children base their decisions on different types of rules.

Problem Type	Rule 1	Rule 2	Rule 3	Rule 4
Balance	100%	100%	100%	100%
Weight	100%	100%	100%	100%
Distance	0%	100%	100%	100%

TABLE 10.1   Different types of problems in the balance beam task and the percentage of correct responses to these problems when children base their decisions on different types of rules.

Problem Type	Rule 1	Rule 2	Rule 3	Rule 4
Conflict-Weight	0%	100%	33%	100%
Conflict-Distance	0%	0%	33%	100%
Conflict-Balance	0%	0%	33%	100%

2. The problems fall exactly into the taxonomy described in Table 10.1. Notice that the child's predicted pattern of responses across the 24 problems does not necessarily improve steadily across all stages. For example, level of performance on the conflict-weight problems deteriorates when the child starts applying Rule 3 instead of Rule 2. The network should be able to capture these regressions in behavior as well as the sequence of stages found in children.

Exercise 10.5

•   The first four patterns and the final four patterns in the test set are those which should yield a balanced beam. The rest of the patterns require that the beam go down on one side. The network has not yet learned how to make the beam go down on one side. It does well on the balanced beam problems by accident—the weights in the network are initially randomized in the range ±0.5 so the effects of input activity along the distance and weight channels in the network are more or less equivalent during early learning. In this respect, the network does not seem to offer a good model of the child's early representation of this problem.

Exercise 10.6

- Since the weights in the network have not had much of an opportunity to adapt to the task, they will still be restricted to values around ±0.5 . This means that the net activity flowing into the hidden and output units in the network will be around zero. Zero input to a unit with the logistic activation function yields an output of 0.5.

Exercise 10.7

1. There is really nothing special about 0.33. McClelland might just as well have chosen 0.5. However, given that there are three target activities for the output units in this network (1.0, 0.5 and 0.0), it makes sense to require that output activations are at least one 'zone' apart when deciding which side of the beam goes down. Notice that this criterion is somewhat loose in that it doesn't require the output units to achieve their target activations to produce a 'correct' response. For example, if the network had produced output responses 0.5 and 0.1 to problem 5 it would be interpreted as correct even though the target activations are 1.0 and 0.0.

2. The output activations for both output units after 401 sweeps (using a random seed of 1—without replacement) is shown in Table 10.2. None of the balance beam problems produce activation differences that exceed criterion. So in this case we'll assume that the network judges the beam to balance in all cases. This is the correct answer for problems 1–4 and

TABLE 10.2   Output activations for the balance beam problem after 401 sweeps, together with the activation differences, target responses and actual responses when judged according to McClelland's 0.33 difference criterion.

	Output Unit Activations		Difference	Response	Target
1.	0.546	0.486	0.06	Balance	Balance
2.	0.647	0.418	0.229	Balance	Balance
3.	0.577	0.396	0.181	Balance	Balance
4.	0.501	0.385	0.116	Balance	Balance

TABLE 10.2 Output activations for the balance beam problem after 401 sweeps, together with the activation differences, target responses and actual responses when judged according to McClelland's 0.33 difference criterion.

	Output Unit Activations		**Difference**	**Response**	**Target**
5.	0.6	0.383	0.217	Balance	Left
6.	0.625	0.413	0.212	Balance	Left
7.	0.484	0.473	0.011	Balance	Right
8.	0.567	0.384	0.183	Balance	Right
9.	0.549	0.499	0.05	Balance	Left
10.	0.49	0.41	0.08	Balance	Left
11.	0.633	0.388	0.245	Balance	Right
12.	0.587	0.375	0.212	Balance	Right
13.	0.665	0.386	0.279	Balance	Left
14.	0.55	0.392	0.158	Balance	Left
15.	0.47	0.426	0.044	Balance	Right
16.	0.527	0.527	0	Balance	Right
17.	0.534	0.466	0.068	Balance	Left
18.	0.472	0.492	-0.02	Balance	Left
19.	0.56	0.312	0.248	Balance	Right
20.	0.668	0.417	0.251	Balance	Right
21.	0.521	0.43	0.091	Balance	Balance
22.	0.468	0.501	-0.033	Balance	Balance
23.	0.539	0.326	0.213	Balance	Balance
24.	0.557	0.345	0.212	Balance	Balance

21–24 but wrong for all the others. Notice that some node activations are quite a long way off the target activations of 0.5 for balanced beams (see problems 2, 4, 23 and 24). If we introduced more stringent criteria that all output activations should be within, say 0.1 of their targets, then only problems 1, 3, 21 and 22 would be judged to balance correctly.

3. On the basis of these results it is rather difficult to classify network performance according to Siegler's stages. On the one hand, the network responds correctly to problems 21–24 which belong to the final stage of

development. On the other hand, the network fails on problems that belong to Siegler's stages 1 and 2, i.e., problems 5–8 and 9–16, respectively. However, we saw earlier in this exercise that successful performance at this stage is really an artifact of the network starting off life with small random weights which produce small net inputs to the output units. If we changed the crierial difference in activation values to, say 0.2, the pattern of results changes but always in the direction of going down to the left. For example, problems 5, 6 and 13 would go down to the left (correct) but problems 11, 12, 19 and 20 would also go down to the left (wrong). Furthermore, problems 2, 23 and 24 would no longer balance. This suggests again that the network has not yet extracted information about the importance of weight and distance. It is important therefore to examine the patterns of response in the network after further training.

4. Consider the performance of the network after it has had an opportunity to process all 1125 training inputs. The output activations for the 24 problems in **bb_test.data** after 1201 sweeps is shown in Table 10.3. The network performs badly on problems 9–12, 17–20 and

TABLE 10.3   Output activations for the balance beam problem after 1201 sweeps, together with the activation differences, target responses and actual responses when judged according to McClelland's 0.33 difference criterion.

	Output Unit Activations		Difference	Response	Target
1.	0.463	0.669	-0.206	Balance	Balance
2.	0.388	0.738	-0.35	Right	Balance
3.	0.377	0.714	-0.337	Right	Balance
4.	0.415	0.648	-0.233	Balance	Balance
5.	0.858	0.208	0.65	Left	Left
6.	0.948	0.102	0.846	Left	Left
7.	0.188	0.866	-0.678	Right	Right
8.	0.209	0.85	-0.641	Right	Right
9.	0.537	0.587	-0.049	Balance	Left
10.	0.374	0.69	-0.316	Balance	Left
11.	0.368	0.744	-0.376	Right	Right

TABLE 10.3   Output activations for the balance beam problem after 1201 sweeps, together with the activation differences, target responses and actual responses when judged according to McClelland's 0.33 difference criterion.

	Output Unit Activations		Difference	Response	Target
12.	0.438	0.65	-0.212	Balance	Right
13.	0.909	0.161	0.748	Left	Left
14.	0.859	0.207	0.652	Left	Left
15.	0.135	0.892	-0.757	Right	Right
16.	0.147	0.901	-0.754	Right	Right
17.	0.202	0.854	-0.652	Right	Left
18.	0.159	0.874	-0.715	Right	Left
19.	0.686	0.378	0.308	Balance	Right
20.	0.884	0.215	0.669	Left	Right
21.	0.193	0.854	-0.661	Right	Balance
22.	0.148	0.884	-0.736	Right	Balance
23.	0.773	0.273	0.5	Left	Balance
24.	0.238	0.812	-0.574	Right	Balance

21–24. It only just misses criterion on problems 2 and 3. Incorrect balance beam judgements are restricted to the *distance, conflict-distance* and *conflict balance* problems (see Table 10.1). This is symptomatic of Rule I behavior. Network performance after 1601 sweeps is shown in Table 10.4. The network can now solve *distance* problems but it is still unable to solve the *conflict-distance* and *conflict balance* problems. Its behavior reflects Rule 2. Finally, consider the behavior of the network

TABLE 10.4   Output activations for the balance beam problem after 1601 sweeps, together with the activation differences, target responses and actual responses when judged according to McClelland's 0.33 difference criterion.

	Output Unit Activations		Difference	Response	Target
1.	0.693	0.277	0.416	Left	Balance
2.	0.366	0.604	-0.238	Balance	Balance

TABLE 10.4   Output activations for the balance beam problem after 1601 sweeps, together with the activation differences, target responses and actual responses when judged according to McClelland's 0.33 difference criterion.

	Output Unit Activations		Difference	Response	Target
3.	0.42	0.515	-0.095	Balance	Balance
4.	0.63	0.303	0.327	Balance	Balance
5.	0.963	0.028	0.935	Left	Left
6.	0.989	0.01	0.979	Left	Left
7.	0.151	0.825	-0.674	Right	Right
8.	0.158	0.814	-0.656	Right	Right
9.	0.86	0.127	0.733	Left	Left
10.	0.638	0.293	0.345	Left	Left
11.	0.274	0.691	-0.417	Right	Right
12.	0.467	0.472	-0.005	Balance	Right
13.	0.972	0.024	0.948	Left	Left
14.	0.96	0.033	0.927	Left	Left
15.	0.103	0.873	-0.77	Right	Right
16.	0.108	0.879	-0.771	Right	Right
17.	0.223	0.747	-0.524	Right	Left
18.	0.161	0.815	-0.654	Right	Left
19.	0.854	0.106	0.748	Left	Right
20.	0.952	0.041	0.911	Left	Right
21.	0.16	0.809	-0.649	Right	Balance
22.	0.153	0.824	-0.671	Right	Balance
23.	0.935	0.045	0.89	Left	Balance
24.	0.206	0.757	-0.551	Right	Balance

after 6001 sweeps as summarized in Table 10.5. Performance has improved markedly on the the *conflict-balance* problems (3 out 4 correct) and improved marginally on the *conflict-distance* problems (1 out of 4 correct). Significantly performance on the *conflict-weight* problems has deteriorated. This pattern of responses is symptomatic of the transi-

TABLE 10.5 Output activations for the balance beam problem after 6001 sweeps, together with the activation differences, target responses and actual responses when judged according to McClelland's 0.33 difference criterion.

	Output Unit Activations		Difference	Response	Target
1.	0.503	0.482	0.021	Balance	Balance
2.	0.379	0.61	-0.231	Balance	Balance
3.	0.39	0.602	-0.212	Balance	Balance
4.	0.453	0.533	-0.08	Balance	Balance
5.	0.97	0.03	0.94	Left	Left
6.	0.992	0.008	0.984	Left	Left
7.	0.011	0.988	-0.977	Right	Right
8.	0.029	0.97	-0.941	Right	Right
9.	0.996	0.005	0.991	Left	Left
10.	0.934	0.064	0.87	Left	Left
11.	0.013	0.987	-0.974	Right	Right
12.	0.153	0.84	-0.687	Right	Right
13.	0.7	0.321	0.379	Left	Left
14.	0.709	0.301	0.408	Left	Left
15.	0.132	0.866	-0.734	Right	Right
16.	0.359	0.642	-0.283	Balance	Right
17.	0.891	0.11	0.781	Left	Left
18.	0.601	0.405	0.196	Balance	Left
19.	0.434	0.574	-0.14	Balance	Right
20.	0.405	0.628	-0.223	Balance	Right
21.	0.386	0.607	-0.221	Balance	Balance
22.	0.589	0.416	0.173	Balance	Balance
23.	0.502	0.502	0	Balance	Balance
24.	0.156	0.843	-0.687	Right	Balance

tion from stage 2 to stage 3 (see Table 10.1). Network behavior in this simulation never gets beyond stage 3. You might like to try running some other seeds to see if you can find a network that reaches stage 4.

Exercise 10.8

- Although the description in the text offers one solution to the balance beam task, it turns out that there are many others—just as there are many solutions to the Boolean functions OR, AND and XOR (see Chapter 3). Figure 10.4 depicts the connection matrix for the balance beam network

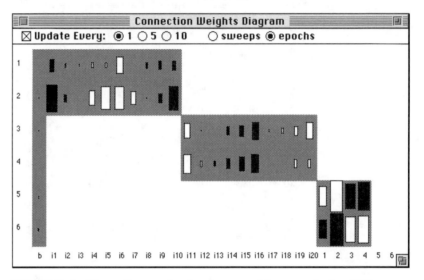

FIGURE 10.4   Hinton diagram showing the state of the network after 6001 sweeps when it is performing according to Siegler's rule 3.

after 6001 sweeps. The biases for hidden and output nodes (column b) are quite small. The connections feeding into the first hidden unit (row 1) from the left-hand weight units are roughly equal and opposite in magnitude to the connections from the right hand weight units. A similar relationship holds for the connections feeding into the second hidden unit from the left and right weight units (row 2). This means that when there are identical weights on either side of the balance beam, their effects will cancel each other out in both hidden units in the weight channel. The same situation holds for the pattern of connections in the distance channel from input units to the hidden units (rows 3 and 4). So although the network has not solved the problem in exactly the fashion described in the text, it is using the same type of solution—let the left and right values on any given dimension cancel each other out. This strategy is maintained in the connections between the hidden units and the output units (row 5 and 6). The connections feeding into the left-hand output unit from the hidden units in the weight channel are roughly equal and opposite in magnitude to the connections feeding into the left-hand output unit from the hidden units in the distance channel. The pattern is repeated again in the connections from the hidden units to the right-hand output unit. Notice that the magnitude of the connections in the *distance* channel are roughly the same as those in the *weight* channel.

Exercise 10.9

- Consider the state of the connection matrix of the network when it is performing according to Siegler's rule 1, shown in Figure 10-5. Clearly,

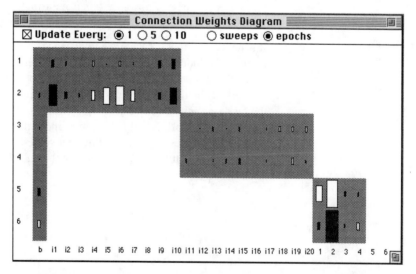

FIGURE 10-5   Hinton diagram depicting the state of the network after 1201 sweeps when it is performing according to Siegler's rule 1.

the magnitude of the connections from the *weight* inputs to the first pair of hidden units is greater than that of the *distance* inputs to the second pair of hidden units. Similarly, the magnitude of the connections from the hidden units in the *weight* channel to the output units is greater than that of the connections from the hidden units in the *distance* channel to output units.

Exercise 10.10

1. McClelland (1989) argues that children have more experience of weight than distance variation in problems involving *torque*. For example, they may pick up objects which vary in weight and appreciate the kinaesthetic differences involved. These experiences will not be accompanied by an equivalent variation in distance from the body—the length of the arm is fixed. The composition of the training set reflects this assumption.

   Notice that it would be just as easy to manipulate the training set so that distance varied more than weight—in which case the network would discover the role of distance before weight. This suggests that there need not be any absolute salience attributed to a particular dimension. The *effective* salience depends on the task being learned.

2. It is not crucial that the weight and distance channels are separated at the input to the hidden unit level. The network *might* learn to segregate the channels spontaneously during training. Imposing the channels on the network at the outset, provides it with an important clue for structuring the task—an innate bias if you will. In the absence of the channel constraint, the network is unlikely to find the same type of solution to the problem. However, you should find that some fully connected networks will still do a pretty good job at solving the balance beam problems.

CHAPTER 11   *Learning the*
English past tense

Theoretical overview

When learning the English past tense, children occasionally produce incorrect forms of the verb like *go-ed* or *hitted* (Berko 1958, Ervin 1964). These errors are usually referred to as *over-regularization* because they seem to reflect children's knowledge about the way other verbs in the language—the regulars—form the past tense. Sometimes, these errors are produced after children have apparently mastered the correct past tense form of the verb. That is to say, children seem to unlearn what they already knew. This observation has often been interpreted as indicating that young children initially acquire the past tense forms of irregular verbs like *go* and *hit* through a process of *rote learning*. However, as they experience more verb forms, they discover the regular pattern for the past tense, i.e., add an /-ed/. In other words, children learn the **rule** for past tense formation in English. They then apply this rule even to forms that do not obey the rule. Further progress is made when children discover that there are exceptions to the rules of English. This requires that children exploit two representational systems—one for dealing with the rule governed forms (the regulars) and one for managing the exceptions (the irregulars). The phenomenon of overregularization thus leads to the proposal that the profile of past tense acquisition implies the existence of *dual mechanisms*.

    In recent years, this view has been challenged by psychologists applying neural network models as a framework for understanding children's linguistic development. For example, Rumelhart & McClelland (1986)—R&M—have shown how a simple feedforward network can be used to model children's acquisition of the English past tense. A simplified version of the model they used is shown in Figure 11.1.

Wickelfeature Representation of Past Tense

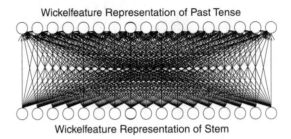

Wickelfeature Representation of Stem

FIGURE 11.1   A feedforward network for learning the English past tense

The network maps a representation of the stem of the verb to its past tense form. Each of the input units is considered to represent some phonetic feature of the verb. Different groups of units represent different phonemes in the verb.[1] Like children, the network must learn to deal with both regular verbs and irregular verbs. So it must learn that the past tense of *grow* is *grew* while the past tense of *show* is *showed*. R&M trained the network on 420 different stem/past tense pairs.

The network was trained using a probabilistic version of the Perceptron Convergence algorithm. When an input verb stem produces an incorrect version of the past tense at the output, the weights in the network are adjusted so as to reduce the error on a subsequent presentation. The same matrix of weights was used by all the verbs on which the network was trained. R&M were able to show that a single network was able to learn simultaneously both the irregular and regular verbs. There was no need for a dual mechanism. Furthermore, it became unclear whether it was appropriate to characterize the network as having learned a rule for the past tense. If the network had learned an add /-ed/ rule for the past tense, then it should have applied the rule to all the stems in the training set—which it did not.

---

1. A phoneme is the smallest unit of speech that can cause a change in the meaning of a sequence of sounds making up a word. Thus, /b/ in front of /at/ makes the word /bat/ while /c/ in front of the same sequence gives the word /cat/. So /b/ and /c/ are distinct phonemes.

R&M also measured the performance of the network while it was being trained on the past tense task. The overall performance on regular and irregular verbs is shown in Figure 11.2. Regular verbs showed

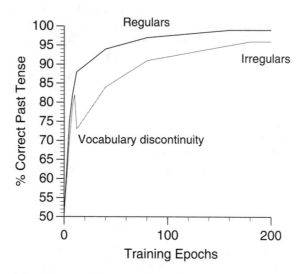

FIGURE 11.2   Network performance on regular and irregular verbs

a gradual increase in correct performance throughout the training of the network. However, the irregular verbs showed a sharp decrement in performance around the tenth training epoch. At this point in training, the network started incorrectly overregularizing some of the irregular verbs like *go* and *hit* to produce *go-ed* and *hitted*. Furthermore, it was at this point in training that the network demonstrated a capacity to generalize to novel forms, i.e., to verb stems that it had never seen before. As you might expect, it had a strong tendency to regularize novel stems, e.g., to output *ricked* when the input is *rick*—a form is hasn't been trained on. The network was successful then not only in succeeding to perform the task of learning the past tense of English verbs but *en route* to doing so, demonstrated some of the same phenomena that are observed in young children acquiring these forms. The R&M model offered a new way for linguists and psychologists to think about language and development.

Unfortunately, there was a major problem with the R&M model— the point at which overregularization errors began coincided with an enforced discontinuity in the verb vocabulary on which the network was trained. Consider again the learning curve for the regular and irregular verbs obtained in the R&M model as shown in Figure 11.2. The dip in performance on irregular verbs comes after 10 epochs of training. In fact, for the first 10 epochs of training, R&M used just 10 stem/past tense pairs. Of these 10 pairings, 8 were irregular. As you can imagine, a network of this size (460 input units and 460 output units) had no problem learning the training set. In fact, there were so many connections available that it could easily learn these 10 input patterns by rote. Indeed, no generalization to novel stems was observed after 10 epochs of training.

At this point in training, however, R&M increased the size of the vocabulary of verbs to 420. Of these 410 new stem/past tense pairs, 334 were regular verbs. Since the network was now trained mostly on regular verb types, the weight matrix changed in a manner which was most suited to regular verbs—though not to irregular verbs. Consequently, the irregular verbs were overregularized. Continued training on the verbs gradually resulted in an improved performance on the irregulars.

R&M justified their tailoring of the training set on the grounds that irregular verbs are generally more frequent in the language than most regular verbs. Furthermore, they noted that there is a vocabulary spurt in children learning language towards the end of their second year. However, children do not start typically to produce overregularization errors until the end of their third year. And the vocabulary spurt observed in the second year results from an increase in the number of *nouns* in their linguistic productions. Thus, the vocabulary discontinuity introduced by R&M does not seem to reflect the conditions under which children are learning the past tense of English. Indeed, Pinker & Prince (1988) have argued that the number of regular and irregular verbs in children's productive vocabularies is relatively balanced during the first three years. Hence, any overregularization errors should not be explained in terms of an imbalance in the sizes of the different verbs types.

More recently, Plunkett & Marchman (1991)—henceforth P&M—have shown how a network architecture using hidden units can reproduce the pattern of errors observed in children, in the absence of any discontinuity in the training set. P&M argued that errors should still occur in the network, just so long as there are competing verb types (regular and irregular) in the training set. Interference between these different types will result in temporary patterns of erroneous performance. No manipulation of vocabulary size should be required.

In this chapter, you will replicate some of the P&M simulations and evaluate the level of performance and pattern of errors produced by a network trained on a task analogous to the English past tense.

Preparing the past tense simulation

The training environment

The first task is to define a training environment for the network. We will assume that the network's task is to learn to map a set of verb stems to their past tense forms. This corresponds to the assumption that children attempt to discover a systematic relationship between the sound patterns that characterize verb stems and their past tense forms.

P&M trained their network on four types of verbs:

1. Regular verbs that add one of the three allomorphs[2] of the /-ed/ morpheme to the stem of the verb to form the past tense, e.g., /pat/⟶/patted/.

2. No change verbs where the past tense form is the same as the stem, e.g. /hit/⟶/hit/. Note that all no change verbs end in an alveolar consonant—a /t/ or a /d/ (Pinker & Prince 1988).

3. Vowel change verbs where the vowel in the stem is changed while the past tense form retains the same consonants as the stem form, e.g., /come/⟶/came/. Note that the vowel change is often conditioned by the final vowel and consonant in the stem (/sing/⟶/sang/, /ring/⟶/rang/). Vowel change stems can thus be considered to form a family resemblance cluster of sub-regularities in the language (Pinker & Prince 1988). A set of sub-regularities that was *omitted* from this model was the

2. The suffix used for the past tense of regular verbs in English depends on the final phoneme in the stem of the verb. If the final phoneme is voiced (as in *arm*) then the [d] allomorph is used (*arm-[d]*). If the final phoneme is unvoiced (as in *wish*) then the [t] allomorph is used (*wish-[t]*). If the final phoneme is an alveolar stop (as in *pit*) then the epenthesized allomorphy is used (*pit-[id]*).

blend family (/sleep/→/slept/, /weep/→/wept/, /creep/→/crept/) where *both* the vowel is altered and a suffix is added. (You can add some to the training set if you wish.)

4.   Arbitrary verbs where there is no apparent relation between the stem and the past tense form, e.g., /go/→/went/.

We will use a similar classification of verbs for our training set.

Input and output representations.

Each phoneme will be represented as a distributed pattern of activity across a 6 bit vector. Each bit in the vector marks the presence or absence of an articulatory feature used to produce the phoneme, such as voiced/unvoiced, back/front, etc. The coding scheme is shown in Table 11.1. You will find a file called **phonemes** in the **tlearn**

TABLE 11.1   Phonological coding used in the Plunkett & Marchman (1991) simulation. Suffix activations are not represented here.

Phoneme	ASCII	Cons/Vow	Voicing	Manner		Place	
		#1	#2	#3	#4	#5	#6
/b/	b	0	1	1	1	1	1
/p/	p	0	0	1	1	1	1
/d/	d	0	1	1	1	1	0
/t/	t	0	0	1	1	1	0
/g/	g	0	1	1	1	0	0
/k/	k	0	0	1	1	0	0
/v/	v	0	1	1	0	1	1
/f/	f	0	0	1	0	1	1
/m/	m	0	1	0	0	1	1
/n/	n	0	1	0	0	1	0
/h/	G	0	1	0	0	0	0
/d/	T	0	0	1	0	1	0
/q/	H	0	1	1	0	1	0
/z/	z	0	1	1	0	0	1
/s/	s	0	0	1	0	0	1
/w/	w	0	1	0	1	1	1
/l/	l	0	1	0	1	1	0
/r/	r	0	1	0	1	0	1

TABLE 11.1   Phonological coding used in the Plunkett & Marchman (1991) simulation. Suffix activations are not represented here.

Phoneme		ASCII	Phonological Feature Units					
			Cons/Vow	Voicing	Manner		Place	
			#1	#2	#3	#4	#5	#6
/y/		y	0	1	0	1	0	0
/h/		h	0	0	0	1	0	0
/i/	(eat)	E	1	1	1	1	1	1
/I/	(bit)	i	1	1	0	0	1	1
/o/	(boat)	O	1	1	1	0	1	1
/^/	(but)	^	1	1	0	1	1	1
/u/	(boot)	U	1	1	1	1	0	1
/U/	(book)	u	1	1	0	0	0	1
/e/	(bait)	A	1	1	1	1	1	0
/e/	(bet)	e	1	1	0	0	1	0
/ai/	(bite)	I	1	1	1	0	0	0
/æ/	(bat)	@	1	1	0	1	0	0
/au/	(cow)	#	1	1	1	1	0	0
/O/	(or)	*	1	1	0	0	0	0

folder containing this coding scheme. ASCII characters are used to represent each phoneme. The ASCII characters W, X, Y and Z are used to represent the absence of a suffix and the three allomorphs of the suffix respectively. Non-phonological codes are used for these past tense endings. They are the 2-bit patterns 0 0, 0 1, 1 0 and 1 1 respectively.

Exercise 11.1

- Why do you think Plunkett & Marchman used a different coding scheme for the suffix?

Create an input set of verb stems

We will simplify the problem by training the network on verb stems that have a constant length—this will make it easier for you to analyze the output from the network and yet still capture the essence of the past tense learning problem. Each verb will consist of three phonemes ordered in a sequence that conforms to the phonotactics (legal sounds sequences) of English. A list of 500 verbs can be found in the file called **stems** in the **tlearn** folder. Use the **Translate...** facility under the **Edit** menu to convert **stems** to a binary representation of the verb stems that obeys the phonological coding scheme stored in the **phonemes** file. Save the resulting file as **phone.data** and format with the appropriate header information for **tlearn**.

Create an output set of past tense forms

Now you need to create a file called **phone.teach** that contains all the corresponding past tense codes for the verb stems in **phone.data**. Before you can do this, you need to decide which class each verb stem belongs to—regular, no change, vowel change or arbitrary. This can be quite a tedious task since class membership depends to some extent on the phonological form of the verb stem. For example, all no change verbs must end in an alveolar consonant—a /t/ or a /d/. Furthermore, you need to select the right allomorph of /ed/ for each regular past tense form (see Footnote 2). So we have performed this task for you. The file **pasts** contains a list of 2 arbitrary past tense forms, 410 regulars, 20 no change verbs and 68 vowel change verbs, in that order. Each line in the **pasts** list is simply the past tense form of the verb stem in the corresponding line of the **stems** list. Convert this list into a binary representation of phonological features of the past tense and save the files as phone.teach.

Exercise 11.2

1. Now construct the **phone.cf** file. Include 30 hidden units in your architecture. How many input and output units should the network have?

2. Do you think you need hidden units to run this simulation?

3. Train the network with a learning rate of 0.3 and a momentum of 0.8 for 50 epochs. Has the network solved the problem? Try a variety of other network parameters if the error level remains high.

Analyzing network performance

It can be difficult to interpret network performance by examining the average RMS error for the whole training set. To guarantee criterial performance on every output unit for every pattern (using the usual rounding off criterion), the average RMS error would need to be less than $\sqrt{(0.5)^2/10000} = 0.005$. It is unlikely that you will achieve this level of error in your simulation.

The most precise way to determine the performance of the network on the past tense task is to examine the output activations on a node by node basis. You can do this by specifying the latest weights file in the **Testing options...** dialogue box and **Verify the network has learned** on the training set. However, when you have 500 patterns to examine and 20 nodes in each pattern, this can quickly become a tedious business! We will now consider various shortcuts to evaluating network performance on this task.

One way to get a quick idea of how the network is performing on individual patterns is to calculate the error for individual patterns in the training set and display the error in the **Error Display**. To do

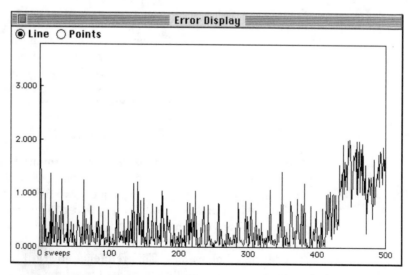

FIGURE 11.3   RMS error for individual verbs in the **phone1** project after 50 epochs of training.

this you must check the **Calculate error** box in the **Testing options...** dialogue box. When you attempt to verify that the network has learned, **tlearn** will display the RMS error for each pattern in the **phone.data** file, in the order that it appears in the file. Since you know the order of the verbs in this file it is relatively easy to determine which verbs are performing poorly and which verbs are performing well. Figure 11.3 displays the individual errors for the **phone** verbs after 50 epochs of training.

It is clear from the error display in Figure 11.3 that many verbs are doing rather well with an RMS error less than 0.5. There are also a number of verbs with an RMS error in excess of 1.0 distributed across the whole training set. Interestingly, though, the verbs with the largest errors are those at the beginning of the training set and those towards the end. Recall that the arbitrary verbs are placed at the beginning of the **phone.data** and **phone.teach** files while the no change verbs and vowel change verbs are placed at the end of these files. Clearly, the irregular verbs are performing the least well in this simulation.

Exercise 11.3

1. Why do you think the irregular verbs are having such a hard time?

2. Examine the output activations for some of the irregular verbs. What is the network doing with them? *Hint: You can use the node activation display to get a good idea of how the input activations are being transformed by the network.*

Using the Output Translation utility

Ideally, we would like to be able to determine the phonemic output of the network from the set of node activations we obtain for past tense forms in the training set. Fortunately, we have a utility program called **Output translation...** in the **Special** menu that enables you do this. Before you can use this utility, you will need to create a reference file that helps **tlearn** translate the real-numbered output vector into a string of phonemes. This reference file will be based on the **phonemes** file you used to translate the original verbs into binary vectors. Make a copy of the **phonemes** file (call it **phonemes.out**) and open the file in **tlearn**. Modify the file so that it is identical to that shown in Figure 11.4.

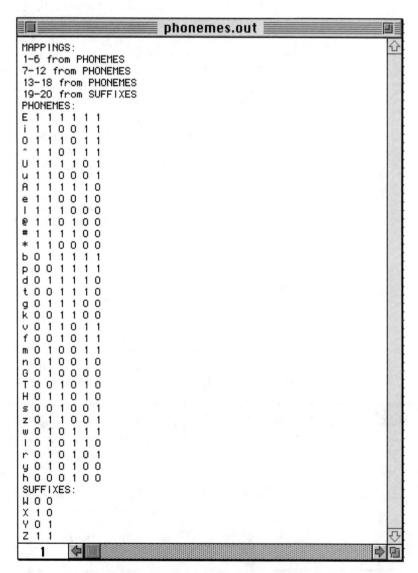

FIGURE 11.4 Output Translation file—**phonemes.out**—for converting the output vector back to an ascii representation of phonemes.

The Output Translation file shown in Figure 11.4 follows a strict format:

MAPPINGS: The first section of the file tells `tlearn` how to divide up the output vector into meaningful chunks. In the current example, the first 6 output units represent a single phoneme, as do the next two sets of 6 output units. Output units 19-20 represent the suffix. The MAPPINGS section of the file is always at the beginning of the file and the line MAPPINGS: is always the first line of the file (remember the colon). Subsequent lines identify groups of units and the codes that will be used to translate them. Hence, the line '1-6 from PHO-NEME' indicates that the first 6 output units will be translated in accordance with the codes in the PHONEMES section of the file. You may only use CODENAMEs that are defined in subsequent sections of the file.

<CODENAME>: Subsequent sections of the file define the binary codes for the ASCII characters. **Output Translation...** will decipher the network's ASCII output by referring to these codes. You can define as many sections as you wish but the binary vectors within a section must all be of the same length. In Figure 11.4, it is important not to mix up the PHONEMES definitions with SUFFIXES definitions. Make sure that the number of elements in the binary vector agrees with the number of output units you have assigned to this CODENAME in the MAPPINGS section. The CODENAME must be in uppercase letters (numbers are not permitted) and end with a colon.

When you have created and saved the reference file, choose **Output Translation...** from the **Special** menu. The dialogue box

Dissecting Euclidean Distance and Threshold Output

Euclidean Distance:

You can choose between two output mapping criteria in the Output Translation utility. The Euclidean distance option determines the closest legal vector, as defined in the reference file, to the current output. No account is taken of the absolute measure of proximity to the closest legal output. In other words, this option "forces the network to talk".

Threshold Output:

All output activations are rounded off to binary values. If the resulting vector matches a legal vector, as defined in the reference file, then the corresponding ASCII is output. If no match is found, then a "?" is output, indicating that the network doesn't know what to say.

shown in Figure 11.5 will appear. Click on the **Pattern file:** box
and select the **phonemes.out** reference file. Check the **Euclidean
distance** box and then **Apply** the translation. Finally, check the
Use Output Translation box in the **Testing options** dialogue
box. Now when you attempt to **Verify the network has
learned**, the **Output** display will include an ASCII representation
of the output units together with the target output. The output display
will continue to include this ASCII output until you uncheck the **Use
Output Translation** box in the **Testing options** dialogue box.

Exercise 11.4

- Use the **Output Translation** technique to
  assess the performance of the network on the training items. Can you *classify* the type of errors that the
  network makes?

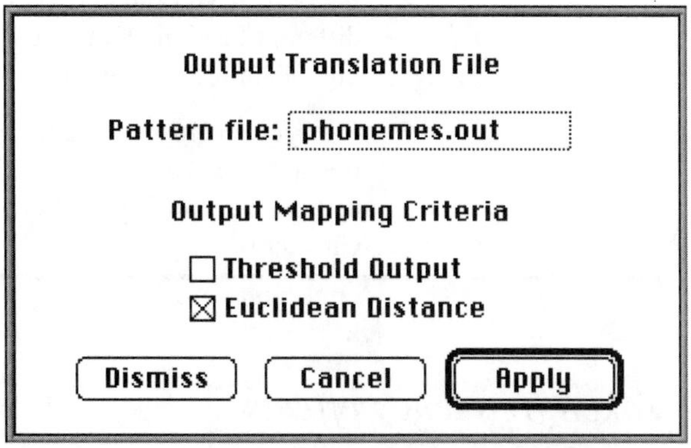

FIGURE 11.5   **Output Translation...** dialogue box

How does the network generalize?

When children and adults are presented with novel verb forms they
are able to generalize their knowledge from other verbs to produce
sensible responses. For example, Berko (1958) elicited past tenses of
novel verbs like *rick* from children. They often responded with

answers like *ricked*. Does your past tense network generalize to novel verbs in a similar fashion? Attempt to determine what generalization your network has discovered about the training set by testing it on verb stems that it has never seen before.

Exercise 11.5

1. In the **tlearn** folder you will find a file called **phone.test** that contains a list of verb stems that the network has not been trained on. Use this list (or part of it) to evaluate how the network generalizes to novel forms. *Hint: Test the network on the set of novel verbs you have selected and use the* **Output Translation** *utility to examine the past tenses produced by the network for inputs it has never seen before.* Does the network generalize equally well to all the verb stems in your test set?

2. If not, can you predict when it will generalize in a different fashion? Evaluate your hypothesis by testing the network on additional novel verbs.

3. Can you decipher the internal representations that the network has built to solve this problem? Try using the analysis techniques introduced in Chapter 6.

Token frequency effects

Although a large proportion of the regular verbs in the previous simulation were learned rather well by the network, the irregular verbs performed badly. Is there any way to improve the performance on the irregular verbs without effecting performance on the regulars? In general, the answer to this question is "no". Anything we do to the irregular verbs will effect the regular verbs since they are both represented by the same weight matrix. However, there is a way to improve performance on the irregulars so that *in the long run* the regulars can still be learned by the network. The solution is simple—just increase the frequency with which the network is exposed to individual irregular verb stems on any training epoch.

Exercise 11.6

- Why should an increase in token frequency of irregular verbs stems lead to an improved overall performance in the network?

You can introduce token frequency effects into your simulation by reproducing each verb stem in your **.data** file and each corresponding past tense verb in your **.teach**. So if there are two copies of each arbitrary verb for one copy of every other verb, then the network will process arbitrary verbs twice as frequently on each training epoch.

Exercise 11.7

1. Modify your past tense training files to increase the frequency of the arbitrary verbs and train the network again. Remember to train the network randomly (with or without replacement). If you train the network sequentially, it will end up processing the same verb many times before it sees any other verbs. This can be detrimental to network performance. Has performance on the arbitrary verbs improved?

2. If you discover that performance improves (if you don't, try manipulating the training parameters in the network), start experimenting with different token frequencies for the other irregular verbs in the training sample. Can you achieve a configuration of verb frequencies that leads to a high level of performance across all the verbs in the training set?

3. When you have found a network and training set that produces good results, determine the generalization properties of the network. Is the generalization different from the first simulation?

Exercise 11.7

---

4. Does network performance and generalization alter with the number hidden units in the network?

U-shaped learning

When you have discovered a training set that enables the network to achieve final successful performance across all verb classes simultaneously, examine the developmental profile of the network. Does the network produce errors on verbs after it has already produced them correctly? Are all verbs equally prone to overregularization? What happens to the regular verbs?

The easiest way to answer these questions is to investigate the performance of individual verbs at different stages of training. The general strategy you will adopt is to re-run your successful simulation, but this time saving the weights at regular intervals. This provides you with "snapshots" of the network at different points in training. You can use these snapshots to evaluate performance on individual verbs and hence determine whether they undergo U-shaped development.

Establishing a developmental profile

Run your successful simulation again but this time save the weights on every training epoch: In the **Training options...** dialogue box, check the **Dump weights every:** box and set the value of **Sweeps** to the number of patterns in your training set. **tlearn** will save a weights file after every epoch, i.e., after it has seen every pattern in the training set. This process creates a lot of files in your current folder so make sure you've got plenty of disk space available!

Next create a test file containing a set of verbs which you wish to evaluate for U-shaped behavior. For example, you may want to consider the arbitrary verbs in the training set. In the **Testing options...** dialogue box, click the **Earlier one:** radio button and select the first weights file that you have just created. Also click on the **Novel data:** radio button and select the **data** file containing your test verbs. You may also wish to send network output to a file, in which case check the **Append output to File:** box and specify a filename. All output will now be saved in a file which you can later edit. You probably also want to use the **Output Translation** utility, so follow the instructions described under "Analyzing network performance" earlier in this chapter.

When you **Verify the network has learned**, `tlearn` will save the ASCII interpretation of the output node activations in the specified output file, as well as displaying these values in the output window. Repeat this whole process by testing the network on consecutive weight files. The output will be appended to the original file you created for the first weights file. You now have a profile of performance for selected verbs during the training of the network.

Exercise 11.8

1. Can you find any examples of U-shaped development?

2. Do errors tend to occur during a particular period of training in the network?

3. Are some verbs more susceptible to errors than other verbs? If so, can you characterize the pattern of behavior?

4. Do you think that the training regime that you have created is a realistic model of the child's learning environment? If not, try creating a new training environment and evaluate performance in the ways you have learned about in this chapter.

Answers to exercises

Exercise 11.1

- There are no compelling reasons to prefer a phonological coding over a non-phonological coding of the suffix or *vice versa*. A phonological coding of the suffix would produce a more differentiated representation of regular and irregular verbs at the output, making it easier for the network to distinguish these two types of verb. On the other hand, a phonological coding would introduce more output units—6 units to represent the [t] and [d] allomorphs and 12 units to represent the epenthesized form. Since the precise form of the suffix was not the main focus of interest for Plunkett & Marchman, they opted for the simpler approach of using just 2 output units to represent potential suffixation processes—the 3 allomorphs or no suffix. Note, however, that the choice of 0 0 to represent no suffix and the other 3 binary patterns to represent the suffixes retains some structure in the suffix versus no suffix conditions. In particular, suffix patterns are linearly separable from the the non-suffix pattern. As a later exercise, you might like to try running the Plunkett & Marchman model with phonological representations for the suffix.

Exercise 11.2

1. Your configuration file should specify that there are 18 input units (6 units for each of 3 phonemes) and 20 output units (18 units to code the stem or modified form of the stem plus 2 units to code the suffix).

2. In order to decide whether the simulation requires hidden units to solve this problem, you need to know whether the problem is linearly separable. Unfortunately, there is no simple way to determine the precise nature of the problem. The easiest way to decide whether the past tense task is linearly separable is to run it with a single-layered network (like Rumelhart & McClelland) and see if it can solve the problem. Remember that single-layered networks using the Perceptron Convergence Rule guarantee to find a solution if a solution exists. Try running the past tense task without any hidden units. You will discover that it fails to pro-

duce the correct past tense forms for all verb stems, though it does a good job on the overwhelming majority. The past tense task is linearly inseparable. It requires hidden units.

3. You should find that the RMS error reduces very quickly during the first few epochs of training. However, the final level of error after 50 epochs varies considerably depending on the initial random seed. If you have difficulty finding a seed which yields a low RMS error after 50 epochs trying using a random seed of 1, trained randomly without replacement. Even with a low RMS error, it can be difficult to determine whether the network has solved the problem from the global error score alone.

Exercise 11.3

1. Most of the verbs in the training set are regular (410 regulars versus 90 irregulars). The regular verb stems attempt to turn on the suffix units at the output while the irregular verbs attempt to keep them turned off. The regular verbs win this competition through shear weight of numbers. The irregular verbs turn on the suffix units too. The network has failed to find a configuration of the weight matrix that allows the irregular and regular verbs to coexist in peaceful harmony.

2. It is easy to see this happening in the network. Use the **Node Acti-vations Display** to display the activity of the output units in response to different verbs stems. You will discover that many of the irregular verbs have one or both of their suffix units switched on. For example, Figure 11.6 shows the pattern of activation produced in the network when the stem of the second training pattern (an arbitrary verb) is presented. One of the suffix units has been switched on inappropri-ately. If you scan through the other irregular verbs in the training set, you will notice that many of them have one or more of their suffix units turned on too.

Exercise 11.4

• Using nearest Euclidean distance as the criterion of correct performance, the network (using a random seed of 1, without replacement after 50 epochs of training) gets 400 (80%) of the training patterns right. Of the

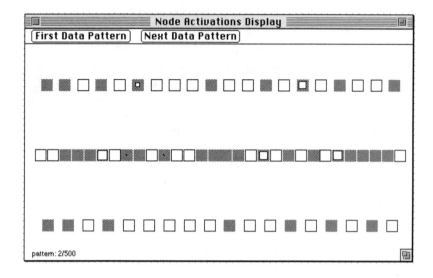

FIGURE 11.6   Node activation display after the network has been trained for 50 epochs. The display shows all the units in the network when the second training pattern is presented to the network. This pattern is an arbitrary verb which has one of its suffix units activated.

410 regular verbs it gets 377 (92%) correct. Of the 90 irregular verbs, it gets 23 (26%) correct. Our rough estimate of verb performance based on Figure 11.3 wasn't too bad! Table 11.2 provides a complete listing of all the errors made on the irregular verbs at the end of training. Error classifications include:

Unclass: The error is unclassified. The output phonemes do not conform to the phonotactics of English or there is no obvious relationship between the input verb stem and the output past tense form. There are only 3 responses of this type—both the Arbitrary verbs and 1 Vowel Change verb.

Overreg: The irregular verb is over-regularized, i.e., the output is the stem of the verb plus a suffix. All the errors on No Change verbs are of this type. 13 of the errors (24%) on Vowel Change verbs are of this type.

Blend: The output consists of a vowel change on the stem and a suffix. These errors are restricted to Vowel Change verbs. There are 41 blend errors—75% of the errors on Vowel Change verbs. Note that there are no Blends specified in the **teach** file. The network has created this response type itself. Children also make this type of mistake.

TABLE 11.2 Classification of errors on the irregular verbs in the **phone** project after 50 epochs of training. See text for definition of error types.

Output	Target	Error	Output	Target	Error
T G H W	t w e W	Unclass	t O r X	t O r W	Blend
T O r X	t A f W	Unclass	m # s X	m A s W	Blend
t @ t Y	t @ t W	Overreg	H E t Y	H e t W	Overreg
p U t Y	p U t W	Overreg	y A r X	y O r W	Overreg
T * t Y	T * t W	Overreg	v # s X	v A s W	Blend
h u t Y	h u t W	Overreg	w * s X	w * s W	Blend
H * t Y	H * t W	Overreg	k O r X	k O r W	Blend
t * t Y	t * t W	Overreg	e t s X	e t s W	Blend
e n t Y	e n t W	Overreg	l O r X	l O r W	Blend
m * t Y	m * t W	Overreg	I g d W	O g d W	Overreg
b i t Y	b i t W	Overreg	w A z X	w A z W	Blend
v A t Y	v A t W	Overreg	b O m X	b O m W	Blend
* l p Z	* l p W	Overreg	v O r X	v O r W	Blend
h E m X	h O m W	Overreg	z E m X	z O m W	Overreg
I s p W	A s p W	Overreg	l A s X	l A s W	Blend
d O m X	d O m W	Blend	H g T W	O g T W	Unclass
b O g X	b O g W	Blend	l O r X	l O r W	Blend
m U l W	m * l W	Overreg	O m z X	O m z W	Blend
y A s X	y A s W	Blend	d A s X	d A s W	Blend
b E f X	b U f W	Blend	r A z X	r A z W	Blend
d E f X	d U f W	Blend	t A r X	t O r W	Overreg
p E f X	p U f W	Blend	T E m X	T O m W	Overreg
f O r X	f O r W	Blend	z O r X	z O r W	Blend
w E t Y	w e t W	Overreg	A r k W	O r k W	Overreg
n U f X	n U f W	Blend	k A s X	k A s W	Blend
g U s Z	g * s W	Overreg	T O g X	T O g W	Blend
p i t Y	p e t W	Blend	p I s X	p * s W	Blend
f A z X	f A z W	Blend	z E f X	z U f W	Blend

TABLE 11.2 Classification of errors on the irregular verbs in the **phone** project after 50 epochs of training. See text for definition of error types.

Output	Target	Error	Output	Target	Error
g O r X	g O r W	Blend	w O m X	w O m W	Blend
l E f X	l U f W	Blend	g I l W	g * l W	Blend
h U f X	h U f W	Blend	r @ s X	r * s W	Blend
z A t Y	z e t W	Blend	k i t Y	k e t W	Blend
m * s X	m * s W	Blend	m O r X	m O r W	Blend
m A z X	m A z W	Blend			

Table 11.3 provides a complete listing of all the errors made on the regular verbs at the end of training. Additional error classifications include:

Identity: The verb is unchanged from the stem to the past tense form. There are 13 errors of this type (40% of the regular verb errors). 12 of these errors occur on verbs that end in an alveolar consonant (a /t/ or /d/) —the ending that characterizes irregular No Change verbs.

Suffix: The network has appended the wrong form of the suffix to the stem of the verb when forming the past tense. There are 11 (33%) errors of this type.

The remaining errors on the regular verbs include 8 blends and 1 unclassified output.

The error patterns on irregular verbs demonstrates the network's tendency to regularize verbs on which it has been trained. However, the error patterns on regular verbs demonstrates that the tendency to regularize is not absolute since some regular verbs are treated as though they are irregular. These are often called *irregularization* errors (see Plunkett & Marchman 1991).

TABLE 11.3 Classification of errors on the regular verbs in the **phone** project after 50 epochs of training. See text for definition of error types.

Output	Target	Error	Output	Target	Error
h E v X	h O v X	Blend	n I f X	n I f Z	Suffix
n E r X	n A r X	Blend	e k s X	e k s Z	Suffix
d u d W	d u d Y	Identity	p A d W	p A d Y	Identity
O r t W	O r t Y	Identity	g u k X	g u k Z	Suffix

TABLE 11.3   Classification of errors on the regular verbs in the **phone** project after 50 epochs of training. See text for definition of error types.

Output	Target	Error	Output	Target	Error
s O s X	s O s Z	Suffix	b e k Z	b * k Z	Blend
E m f X	E m f Z	Suffix	w E f X	w O f Z	Blend
w # s X	w # s Z	Suffix	E r s X	E r s Z	Suffix
l i k Z	l u k Z	Blend	k w k Z	k ^ k Z	Unclass
e s t W	e s t Y	Identity	y E f X	y O f Z	Blend
i k t W	i k t Y	Identity	A s p W	A s p Z	Identity
i k s X	i k s Z	Suffix	w # d W	w # d Y	Identity
# k t W	# k t Y	Identity	H I s X	H I s Z	Suffix
f I s X	f I s Z	Suffix	w U t W	w U t Y	Identity
A r d W	A r d Y	Identity	n * d W	n * d Y	Identity
z I s X	z I s Z	Suffix	d E v X	d O v X	Blend
m i r X	m e r X	Blend	H U d W	H U d Y	Identity
b A d W	b A d Y	Identity			

Exercise 11.5

1. The network does not generalize equally well to all novel stems. Of the 211 novel stems in the **phone.test** file, it regularizes 162 (77%) of them. 12 of the remaining stems produce unclassifiable responses, 8 produce No Change responses, 7 produce pure Vowel Change responses and 22 produce Blends.

2. All the No Change responses are made to novel verb stems that end with an alveolar consonant—a property they share with the No Change verbs in the training set. The majority (20 out of 29) of Vowel Change and Blend responses involve changes to the stem vowel that are characteristic of Vowel Change verbs in the training set.

3. Figure 11.7 shows a cluster analysis of the hidden unit activations when 16 verbs stems from the training set are presented to the network. The 16 verbs include 8 regular verbs, 4 No Change verbs and 4 Vowel Change verbs all of which are correctly inflected by the network. The cluster

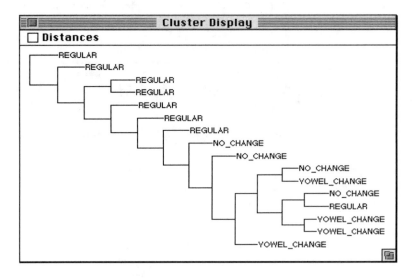

FIGURE 11.7   Cluster analysis of the hidden unit activations for 16 verbs selected from the training set which the network learned correctly. The irregular verbs tend to cluster together in a similar region of hidden unit space whereas the regular verbs are more distributed.

analysis reveals that the irregular verbs are all grouped together on one sub-branch of the tree while most of the regular verbs (7 out of 8) appear on different branches. The analysis shows that irregular verbs have similar internal representations at the hidden unit level whereas the regular verbs are more widely distributed in the hidden unit space. Note, however, that there is no guarantee that the internal representations of regular and irregular verbs will always cluster in this fashion. You will need to check whether your network clustered the verbs this way.

Exercise 11.6

- An increase in the token frequency of irregular verbs results in an increase in the number of times the network processes individual irregular verb stems compared to individual regular verb stems. This provides the network with a greater number of opportunities to reduce the output error for any given irregular stem within a training epoch. Output errors are reduced by changing the weights associated with active connections (connections whose source nodes are turned on). If these weights are also active when processing regular verbs, then performance on regular verbs is liable to deteriorate. However, the high *type frequency* (the number of different verbs that undergo a particular type of transformation, e.g., add a suffix) of the regular verbs tends to protect them from too much interference. The high *token* frequency of the irregular verbs facilitates learning of the irregulars but minimizes interference with the regulars since associated weight changes are specific to a unique input (the irregular form) rather than spread across a variety of inputs—as is the case with regular training.

Exercise 11.7

1. Performance on the Arbitrary verbs will improve as you increase their frequency in the training set. However, you may have to increase the token frequency substantially (say 10 to 15 times the regulars) before the network gets the Arbitrary verbs correct. Note that one of the Arbitrary verbs may require a higher token frequency than the other to achieve success. Note also that the improvement in performance on the the Arbitraries is achieved without any appreciable decrement in performance on the rest of the training set.

2. Increasing the token frequency of the Arbitrary verbs does not result in improved performance on the No Change and Vowel Change verbs. You will need to increase the token frequencies of these forms as well, to achieve good overall performance. No Change and Vowel Change verbs will need a token frequency that is about 3 to 5 times greater than the regulars. Individual verbs may require higher token frequencies.

3. It is difficult to achieve perfect performance across all verb classes when the training set contains so many irregular verbs. However, you should be able to find a configuration of network parameters and token frequencies that enables the network master at least 90% of all four verb types. The network's response to novel stems will be predominantly to add a suffix. However, you will also discover a small increase in the tendency to irregularize novel stems that have phonological characteristics in

common with irregular verbs in the training set. There will also be an increase in the number of irregularization errors for verbs in the training set.

4. We have already seen that the hidden units are critical to successful mastery of the past tense problem. However, the *number* of hidden units used in the network is important too. If the network is trained with just 2 or 3 hidden units, it will perform extremely badly on this task. If it is trained with a large number of hidden units (say 100), it will learn the past tense forms of verbs in the training set very well but it will fail to generalize well to novel verbs. With a small number of hidden units, the network lacks the necessary representational resources to solve the problem. With too many hidden units, the network learns too much about the details of individual verb stems and fails to recognize the regularities that group them together in different classes. For most problems, there is an optimal range of hidden units which both enables the network to

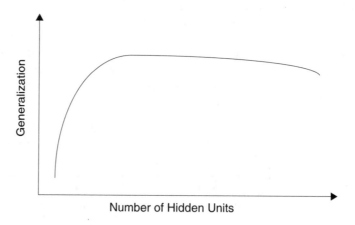

FIGURE 11.8   Occam's cliff

solve the problem and produce good generalization. Figure 11.8 depicts graphically the trade-off that is commonly observed between the number of hidden units and generalization in the network—sometimes referred to as Occam's cliff! 30 hidden units seems to work pretty well on the past tense problem.

Exercise 11.8

1. You should be able to find many examples of U-shaped development both on the regular and irregular verbs. The error types that occur cover the whole range reported earlier—overregularization, blends, identity, suffix and unclassified. The task of sifting through all the data can be rather tedious but it will give you an excellent feel for how the verbs are treated at different stages in training. To simplify the task, you might like to restrict your analysis to a subset of the irregular verbs in the training set. However, a proper evaluation of network performance requires a lot of work—just as it does with young children! Note that U-shaped performance does not occur for all verbs at the same time. Furthermore, the same verb may undergo successive phases of correct and incorrect performance. Plunkett & Marchman (1991) referred to this pattern of behavior as *micro U-shaped development*. Marcus, Ullman, Pinker, Hollander, Rosen & Xu (1992) observed this developmental profile for children learning the English past tense. Figure 11.9 shows the profile of development for the different verb types reported for one of the simulations in Plunkett & Marchman (1991).

2. Most errors occur during the earlier period of training (see Figure 11.9). Of course, this is not surprising since by the end of training the network has mastered most of the verbs in the training set whereas at the beginning of training it knows nothing about the relationship between verb stems and their past tense forms. On the other hand, children tend not to make past tense errors during the earliest stages of learning. This indicates an important failing in Plunkett & Marchman's (1991) attempt to model children's acquisition of the English past tense.

3. Errors are most likely to occur on irregular verbs that have a low token frequency since they will be less robustly represented in the weight matrix of the network. Errors will also be phonologically conditioned. In other words, irregular verbs that sound similar to other irregular verbs may be treated like those irregulars. A common example is a Vowel Change verb that ends in an alveolar consonant being treated as though it were a No Change verb. Regular verbs which resemble irregular verbs may also be irregularized.

4. The training set you have used in this chapter contains 410 regular verbs and 90 irregular verbs. In other words, irregular verbs constitute 22.5% of the training environment. Yet there are only 150–180 commonly used irregular verbs in the entire English language. Clearly, irregular verbs

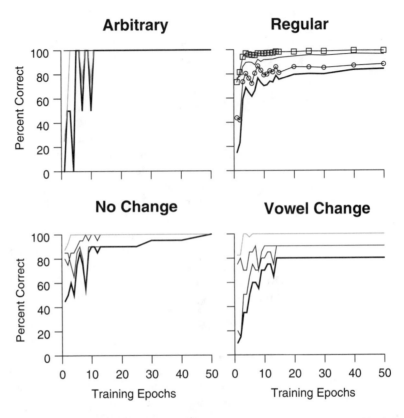

FIGURE 11.9   Performance on regular and irregular verbs in the Plunkett & Marchman (1991) model of the acquisition of the English past tense. The vocabulary size is held constant throughout training. Percentage word hit (solid line, no symbol) reflects overall performance for each verb type. All verb types undergo micro U-shaped development.

are grossly over-represented in the training set. We have, therefore, provided you with another training set in the files **stems2** and **pasts2**. These files contain just 40 irregulars out of a total 500 verbs, i.e., 8% of the training environment. These files also contain repetitions of irregular verbs to capture the token frequency effects discussed above—all irregular verbs are presented with a token frequency of 10 except the Arbitrary verbs which have a token frequency of 20. You will discover that the network does rather well at learning most of the verbs in this training set, though you may have to tweak the token frequencies a bit to get perfect performance. In fact, the simulation reported in Figure 11.9 uses a training environment very similar to **stems2** and **pasts2**. After 50 epochs of training with a random seed of 1 (and no replacement), the network gets 96% of the regulars correct (nearly all the errors are Identity errors on stems that end with an alveolar consonant) and 92.5% of the irregulars correct (3 Vowel Change errors).

As is apparent from Figure 11.9, however, the network still fails to capture the early period of error-free performance that is found in young children. In later modeling work, Plunkett & Marchman (1993, 1996) rectified this problem by training the network using an incremental training regime, i.e., a training set in which the size of the vocabulary starts off small (say 20 verbs) but then gradually increases in size (say by 1 verb every 5 training epochs). The incremental training regime was designed to mimic children's gradually expanding verb vocabularies. This method of training the network succeeded in producing an early period of error-free performance, micro U-shaped learning and even higher levels of overall performance.

It is a simple though laborious process to set up an incremental training regime. You need to define a large number of **data** and **teach** files. Start with a small training environment of say 20 verbs, then create another training environment that contains the same 20 verbs plus one more. Continue in this fashion until your training environment has reached the desired size. Now train the network on the first set of verbs and save the final weights file. Next redefine the training environment (by opening a new project) to increment the verb vocabulary size by one. Continue training but start out with the most recently saved weights file. Repeat this process through all the incremental training environments you have created and then evaluate network performance in the usual fashion.

## CHAPTER 12

The importance of starting small

Introduction

With neural networks—as with humans—it is natural to think that "more is better," in the sense that the more resources a system has, the more powerful it should be. In the case of a network, this might mean more units or weights. In the case of humans, it might mean more neurons, synapses, or more generally, a more mature state of development. So we might disprefer a network with restricted resources, and believe that the immature state of childhood is somehow inferior to the adult state.

Of course, this latter assumption then raises the ticklish question about why so many "higher species" (including primates such as ourselves) have evolved to have such long periods of immaturity. If infancy and childhood are disadvantageous—and certainly, they seem to be, since the immature creature is vulnerable and dependent on the adult caretakers, and the adults' energy and time is often tied up in caretaking—why should evolution not have moved us in the direction of shorter and shorter periods of immaturity? Why should development not have sped up, over evolutionary time, rather than slowing down, as it has seemed to?

We discuss these issues at length in *Rethinking Innateness* of this work (see particularly Chapter 6). One possible answer we consider is that there is an unexpected advantage to having restricted resources. Or, as we put it, there are some cases where it is important to start small. (Newport 1988, 1990, who has proposed a similar hypothesis, calls this the "less is more" hypothesis.)

The example we give below is a simpler version of the simulation we discussed in *Rethinking Innateness*, Chapter 6, which comes from Elman (1993). The problem is as follows.

The problem

Many linguists and psycholinguists have noted that children appear to generalize readily beyond the actual language data they hear and generate novel utterances which sometimes also reflect grammatical structures they have probably not yet heard. So it is clear that children are doing more than simply learning by rote a list of sentences which they call up on demand as needed. Rather, children form some more abstract grammar based on the input they are exposed to.

The space of possible grammars which might be consistent with the limited input available to children, however, is quite large. So we might wonder how children are able to identify just the right grammar. (Of course, it is obvious that children take some time to do so, and early child language is littered with errors which reflect erroneous generalizations. But ultimately, most of us come to settle on what is more or less the correct grammar for our language and dialect.)

In fact, Gold (1967), in a widely cited and highly influential paper, put the problem in somewhat starker terms. He argued that it is theoretically impossible for a language learner to learn the kinds of grammars required for natural languages based solely on positive-only examples. By positive-only examples, Gold meant that the child hears only examples of grammatical sentences, and no examples of the explicit form "'Bunnies is cute' is a bad sentence" (which is termed negative evidence). Gold suggested that this meant that either some other form of negative evidence was indirectly available, or that some aspects of linguistic knowledge were innate. In the latter case, the child can rely on pre-existing knowledge to narrow down the range of possible grammars he or she will induce from the data.

Gold's argument has figured importantly in claims regarding innate linguistic knowledge, although the force of the argument is in fact much weaker than is sometimes realized. For one thing, there are quite likely many sources of indirect negative evidence available to the child (e.g., if in a certain context a child expects to hear a sentence of a certain form, and hears something else instead, the child might infer that there is something wrong with the sentence as he or she anticipated it). Gold's proof also relies on the assumption that the end state is the perfect induction of the completely correct grammar. If one relaxes this assumption slightly and concedes that the final grammar need only be approximately correct (and that grammars differ slightly across individuals), then the proof does not apply. Finally, even if one believes that there are innate constraints on what are possible grammars, the question of what form such constraints take (i.e., how specifically linguistic they are, or whether they are broader constraints having to do with human cognitive and perceptual capacities) is wide open.

Elman's (1993) simulation offers an example of how such a general processing characteristic might have specifically linguistic consequences. Elman's goal in this work was to see if a simple recurrent network could be trained to process sentences with relative clauses, similar to those below.

(1) The dogs ~the cats chase~ bite ferociously.

(2) The dogs ~the cats ~the mouse fears~ chase~ bite ferociously.

The relative clauses are shown as subordinate to the nouns they modify. This notion of subordination (or "part-of" relationship) was seen as a challenge to neural networks, and one that Fodor and Pylyshyn (1988) claimed could not be represented (at least, not unless the network simply instantiated the symbolic solution for representing part/whole or constituent structure). If true, this failing would significantly limit the usefulness of connectionist networks for modeling human cognition. Elman's goal was to see if Fodor and Pylyshyn's claims could be falsified empirically by training a network to process sentences for which constituent structure was crucial.

Elman's initial attempts to train a simple recurrent network in fact failed, despite numerous experiments with different architectures and learning parameters. What did work was a bit surprising. If a network were first trained on a subset of sentences which contained no relative clauses (e.g., simple sentences such as "The dogs bite ferociously"), and then—keeping the modified weights which had been learned for the simple sentences—gradually more and more relative clauses were introduced into the training data, the network was able to learn the complex sentences with relative clauses relatively quickly. Alternatively, a network could be trained from the start on the fully complex sentences, with noise injected into the context units initially after every several words; and then, over time, the frequency of introduced noise decreased until there was no noise at all. This latter manipulation mimicked the effect of having a limited working memory at the start (which allowed only the simple fragments of the complex sentences to be processed), which slowly increased in capacity until the adult level was achieved. Both alternatives, either having an external filter which presented simple sentences first, or having an internal filter which accomplishes the same, led to success on the training set and generalization to novel sentences.

Why does starting small work? The problem for the first networks (the ones that failed) was this. The network had to learn about dependencies between words in sentences which were often widely separated. For instance, in example (2) above, the final verb ("bite") is in the plural form because it goes with the first noun ("dogs"). Determining which verb goes with which noun is not easy. In this case there are two other nouns and two other verbs which intervene, whereas in sentence (1) there are only a single noun and verb which intervene, and in the simple sentence "The dogs bite ferociously" no other words intervene. Learning to develop abstract representations of the sentences which allow the network to learn the correct generalization is not easy. Among other things, the network needs to have good representations of the words themselves (e.g., know which ones are nouns and verbs in the first place—not information which was given explicitly, by the way). These representations are formed on the hidden units and, when they get cycled around to the context layer, form the basis for memory. Lacking good representations means you lack good memory. Unfortunately, it may be difficult to develop good representations in the first place, since the network creates representations based on distributional characteristics which help it identify different grammatical categories of words, but the poor memory makes it difficult to figure out what the relevant distributional characteristics are. There is thus a hard boot-strapping problem.

(This point can easily be made concrete. Imagine listening to an unknown language on the radio, and give yourself the task of repeating what you hear. How well do you think you would do? Probably not well at all. Then imagine how much easier it would be to imitate an English speaker. The benefit of knowing the language means you have available an internal coding which is highly efficient for memory purposes.)

Starting small is one way to solve the boot-strapping problem. By giving the network simple sentences first (or, alternatively, restricting its memory so that it can only process simple fragments), minimal memory is required for the network to infer that some words are nouns, some are verbs, etc., and to represent these in different parts of hidden unit activation space. Once the hidden unit activation space is structured appropriately, the network has representational resources which improve the quality of its memory and make it possible to learn the complex grammatical facts necessary to make sense of the longer sentences. Starting small provides a scaffolding which can be learned easily, and which facilitates learning more complex information later.

This example also suggests a reason why an immature state of development, despite carrying some maladaptive consequences, might nonetheless be selected for by evolution. The network which began in a fully mature state was unable to learn the complex grammar; it was only when resources or input were restricted that networks were successful. Thus immaturity may confer certain important advantages.

Finally, the starting small example illustrates ways in which non-specific constraints may have domain-specific consequences. The networks which are subject to such constraints might well be said to have "innate knowledge of relative clauses," since they and only they were able to master this structure. But it is obvious that the substance of this innate knowledge has nothing specifically to do with relative clauses.

In this chapter you will have an opportunity to explore the starting small example. Because the dataset used in Elman (1993) was large and complex, we have chosen a problem which possesses similar critical features as the relative clause problem, but is easier to work with. Here is the problem.

Simulation

We begin with what is called a Finite State Automaton (or FSA) as shown in Figure 12.1. This is an imaginary device (one could build one, but one never does) which can be used to produce simple sentences. In our case, the sentences will consist not of words but of the letters **A**, **B**, and **X**.

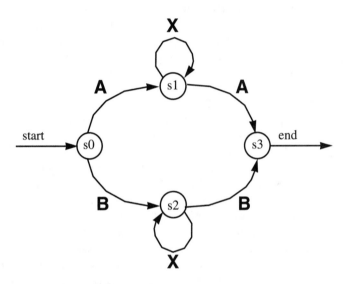

FIGURE 12.1   Finite state automaton used to generate stimuli.

(By the way, although this machine looks like a network, it is quite different. The circles stand for states we can be in rather than processing nodes, and the arcs with arrows represent legitimate paths between states. The machine is a serial machine—we can·be in only one state at a time. Think of this graph as a kind of map which tells us how to proceed from start to end.)

The way we use the FSA is to begin at state s0. We then toss an imaginary coin. If the coin comes up Heads, we move to the top state, s1, and as we do so, we emit the letter on the arc between states. In this case, that letter would be **A**. (If the coin had come up Tails, we would have moved to state s2 and produced a **B**.) We then toss the coin again. This time, if the result is Heads, we proceed to state s3, emitting another **A**, and we're done (and begin with the next "sentence"). If our coin had come up Tails, then we would have taken the link which doubles back to s1 (leaving us in the same state) and produced an **X**. In this case we must keep flipping the coin until eventually we get to state s3 and finish the sentence.

In this way, we might generate the following sorts of strings:

AA	**BXXB**
BXB	**AXA**
AXXXXXXA	**BXXXB**

If we concatenated the above strings together, we would have:

AABXBAXXXXXXABXXBAXABXXXB....

This is a simple "language" which has the following property:

- Sentences begin with either an **A** or a **B**, contain some number (possibly nil) of **X**'s, and end with the same letter with which they begin (**A** or **B**).

Note that we can allow ourselves the option of using "coins" which are weighted in various ways when we are in states s1 or s2, so that the probability of cycling back to the same state (and generating another **X**) can vary. At one extreme we might never generate an **X** so that our sentences will all be of the form **AA** or **BB**. At the other extreme, we might have long sequences of **X**s. A perfectly weighted "coin" would cycle us around with probability 0.5.

This language, although vastly simpler than one which contains relative clauses, shares the characteristic that there are dependencies between elements (the initial and final **A**s and **B**s) which can be interrupted by a strings of **X**s. The problem for the network is that if it gets stuck listening to a long string of **X**s, it may forget what state it is in (s1 or s2) and therefore fail to anticipate the correct end letter, **A** or **B**. Let us first see that this is indeed the case, by "starting big."

Starting big (and failing)

We have supplied you with four data sets for this problem. These data sets are contained in the folder **Chapter12**. They all end with the suffix **.strings**, and have prefixes **ss.0, ss.1, ss.3,** and **ss.5**. The **ss** stands for "starting small", and the numbers (.0, .1, .3, and .5) refer to the probability of producing an **X**, when we're in either s1 or s2. Thus, the file **ss.0.strings** has strings in which there is 0 probability of cycling back to s1 or s2, and so there are no **X**s. Files **ss.1.strings, ss.3.strings**, and **ss.3.strings** were generated with FSAs which were identical except that the probability of cycling back was 0.1, 0.3, and 0.5, respectively. (Note that **ss.5.strings** is the normal case of a fair "coin.") We will start big by training on the **ss.5.strings** data set.

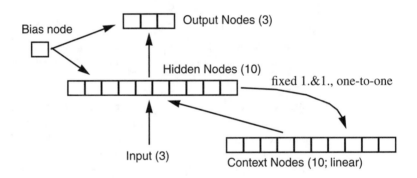

FIGURE 12.2   Simple recurrent network used in this simulation.

File configuration

For this simulation you will need to create a simple recurrent network with 3 input units, 10 hidden units, 3 output units, 10 context units. The 3 input units are required because we have 3 different possible letters (**A, B,** or **X**), and the network's task will be to predict what the next letter will be (again, **A, B,** or **X**). The network as it really should be drawn is shown below in Figure 12.2; the network as it will appear in **tlearn** in the **Network Architecture** graphic is shown in Figure 12.3 Context units and hidden units are placed side-by-side.

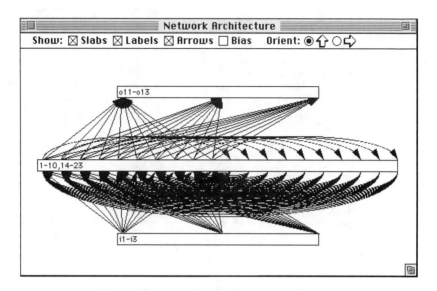

FIGURE 12.3   Simple recurrent network used in this simulation, as it appears in the Network Architecture display.

.cf *file:* Because this dataset has equiprobable transitions (p=0.5) from s1 to s1 and s3 to s3, use the name **ss.5** for your project (i.e., select **New Project** from the **Network** menu). (The .5 refers to 0.50 probability of staying in the same state.) Create the file **ss.5.cf**, following the same directions as for **srn.cf** in Chapter 8, except there will be 10 hidden units (numbered 1-10), 3 output units (11-13), and 10 context units (14-23). The *weight_limit* should be 0.1, and be sure to define the copy-back connections correctly and to make the context units *linear* (refer to Figure 8.3 for details).

.data *and* **.teach** *files:* The dataset you will be using (contained **ss.5.strings**) is not yet in a form to be used directly in the simulation, since it consists of alphabet letters (e.g., **BBBBAABXXBBXB**...., each letter on a line by itself). You will need to convert this file to a file consisting of binary vectors. Since there are 3 different letters, you can use the vector **1 0 0** to represent **A**, **0 1 0** for **B**, and **0 0 1** for **X**. Follow the same procedure used in Chapter 8 to create a **codes** file, and then translate **ss.5.strings**; save the output in **ss.5.data** and then add the header (the **distributed** keyword, and **15008**, which is how many patterns there are). Then

Why are context units numbered last?

The convention of numbering context units last is very important. The number of a node in the network determines the order in which it is updated. Thus, in feedforward networks, we always want our output units to have higher numbers.

In the case of context units, it is important that at the time the units they feed into are updated, the context units themselves still hold the values of the hidden units from the previous sweep. Only after the current sweep has been completely carried out do we want to save the (newly calculated) values of the hidden units in the context units so that they can be available for the next sweep.

The way we accomplish this is to use the following rule: **always assign context units higher numbers than the nodes into which they feed.** For most purposes, it suffices to simply number them last. (As a result, the output units will have lower numbers than the context units.) There may be cases—such as when we have context units feeding into context units—in which the number gets more complex. But the rule above always works.

copy `ss.5.data` to `ss.5.teach`, moving the first pattern to last position.

Training and testing the network

Set the learning rate to 0.1, the number of sweeps to 50,000, the seed to 1 (or any other number you want; just remember it in case you want to replicate your results or try different seeds), and save the error every 1,000 sweeps. Open up the **Error Display** and train the network.

- To verify how well the network learned, create a test file called **test1.data** and containing (concatenated), the 4 representative strings **AA**, **BB**, **AXXA**, and **BXXB**; it will be useful to precede the test set with an additional copy of the strings **AA** and **BB**. (Of course, you will have to use the binary vectors corresponding to these patterns. There are 16 letters total, thus the file will have 16 input patterns, each on a line by itself.) Test the network (using `ss.5.50000.wts`), ignoring performance on the first four inputs (**AA BB**), looking only at the final 12.

Before deciding how well the network has performed, consider what would constitute good performance.

You will probably find that the network performs somewhat reasonably on the sequences with no **X**s (though not perfectly, and possibly not well at all), but not well at all on sequences with embedded **X**s. Or the network may do well with **AXXA** but not **BXXA** (or the reverse).

Why do we need the extra patterns at the beginning of the test data?

During training, the network's context units were never reset (as we did in Chapter 9.) The network simply had to infer when a string began or ended based on what it learned from the training data. Thus, during the bulk of processing, the context units had some value which reflected prior inputs, and the network was able to learn when that context was important (e.g., in the middle of a string), and when it was irrelevant (e.g., at the end of a string).

But there is one special occasion when the context units are unique; that is when the network is started up and gets the very first input. At this time, the context units are still at their default (0.0) state. This only occurs once per run, however, and because the network so rarely encounters "zero" contexts, it doesn't get a chance to learn how to deal with them.

When we test the network, we provide an additional several inputs which just serve the purpose of getting things up and running, and establishing a context which will be more similar to what the network encountered during training. The response to these first few inputs is often somewhat degraded and not representative of later responses, so we ignore them.

Exercise 12.1

1. When predicting the first letter of a string, what should the network output?

2. After the network has gotten the first letter, what should it predict?

3. What should the network predict when it gets **X**s?

4. How can you tell that the network knows it has gotten to the end of the current string? What will its prediction of the next input be?

5. Now, look at the actual output you got. How does the network fare, given your expectations? Does it predict the correct thing at the beginning of each string?

6. Is performance equally good/bad on all types of patterns? Compare the simple ones (**AA**, **BB**) with **AXXA** and **BXXB**.

The results of the simulation are thus similar to the failure of the Elman (1993) network which attempted to deal with all the data at once.

Before proceeding, save the weights for this network which failed by renaming (or moving) the file to one called **ss5-bad.wts**. We will use this file later in this chapter when we discuss Principal Components Analysis.

Starting small

We will now try to train a network in an incremental fashion. We proceed much as before (you should keep the **ss.5** files, since our goal is to be able to process those ultimately), but this time working with the datasets contained in **ss.0.strings**, **ss.1.strings**, and **ss.3.strings**.

For each dataset, create a project file (**ss.0**, **ss.1**, **ss.3**) and associated **.data** and **.teach** files, just as you did for **ss.5**. (Your machine may run out of memory if you try to keep all these files open at once, so close each project's files once they're complete.)

(i) Now open project **ss.0**. This is a dataset with no **X**'s. Set the training options so that the learning rate is 0.1, momentum is 0.0, pick a random seed, dump the error every 1,000 sweeps, and then train for 50,000 sweeps. This will give you a weights file called **ss.0.50000.wts**.

(ii) Next, open project **ss.1**. This is a dataset in which the probability of cycling back from s0 to s0 (and s1 to s1) is 0.1; there are thus a few **X**'s but very few.

Set the training options so that we use the **ss.0.50000.wts** that was just created, and lower the learning rate to 0.08. (This process of gradually lowering the learning rate during training is called "annealing". We do it here because we do not want to move too far from the initial weights we just learned. Rather, we want to fine-tune them.) Train the network for another 50,000 sweeps. This will give you a weights file called **ss.1.100000.wts**.

(iii) Open project **ss.3**. Using the weights file created in the previous step, lower the learning rate once more to 0.05 and train for 50,000 sweeps. This produces **ss.3.150000.wts**.

(iv) Finally, open project **ss.5**. Load in the weights file **ss.3.150000.wts**, lower the learning rate to 0.04, and train for a final 50,000 sweeps. We now have the network we wanted: **ss.5.200000.wts**.

Exercise 12.2

- Consider the same questions you answered in the previous ("Starting big") simulation, and compare the results with those you obtain with the incrementally trained network. How does the performance differ in the two cases? (Hopefully, the second network performs better!)

Why does it work?

You should have found that when you tried to train the network starting with the dataset that contains many embeddings (**ss.5**), the result either does not work, or does not work nearly as well as when you train incrementally, beginning with strings that contain no embedded **X**s. (If you didn't get this result, it would be worth trying again several times with different random seeds. As with human subjects, networks display some variability, and to be confident of our conclusions, we should run a number of replications with different networks and then test the significance of our results with statistical analysis.)

The hypothesis we advanced earlier, in discussing the simulation Elman ran which attempted to train a network to process complex sentences, was that starting small allowed the network to learn the basic internal representations which could then be utilized to learn more complex grammatical facts. A similar account holds for the task here.

Let's look at the internal representations (hidden unit activations) in a network that failed (i.e., a network which "started big," with all the data). To do this, however, we first need to introduce a new topic and tool: Principal Component Analysis.

Using Principal Component Analysis to analyze networks

We have seen repeatedly that insight into a network's behavior can be gained by examining the internal representations which the network forms in the service of carrying out some task. Often, the input representations are of a form which does not permit the network to produce the correct outputs directly; thus, in the XOR task, we have the problem that groups of inputs which are maximally dissimilar in their form (i.e., 00 and 11 on the one hand, and 01 and 10 on the other) must be treated as similar for purposes of the response they produce (in this case, 0 for the first two and 1 for the second two). In order to solve this problem, the network requires a hidden layer on which it re-represents the inputs in a manner which restructures the similarity relations so as to facilitate the correct response.

As networks grow in their complexity and the number of hidden units increases, it is typically difficult or impossible to analyze the similarity structure of the hidden unit patterns by direct inspection. To get around this problem, we have introduced hierarchical clustering as a technique for measuring the distances between hidden unit patterns (because in these networks, distance and similarity are directly related). There is a drawback to hierarchical clustering, however, which can be seen in the following example.

Consider the hypothetical case where we have a network with 3 hidden units, and we have 4 different hidden unit activation patterns we wish to compare. Because there are only 3 hidden units, we might visualize these directly, as shown in Figure 12.4 (a). Let us imagine that these patterns are the activation patterns which are produced when the network processes the word "girl" and "boy" in two different contexts: As the subject of a sentence, and as the direct object.

If we were to carry out a hierarchical clustering of these patterns, we might see something similar to what is shown in Figure 12.5. As we would have expected from what we can see of the spatial relations shown in Figure 12.5 (a), the two patterns corresponding to "boy" are placed on a branch, and the patterns corresponding to "girl" are placed on a second branch.

Notice, however, that in addition to the "boy/girl" distinction which is encoded by the way the hidden unit space is partitioned, there is a second dimension of interest. In Figure 12.4 (a) we can see that when a word is used as the subject of a sentence (whether that word is "boy" or "girl"), the hidden unit representation lies closer to

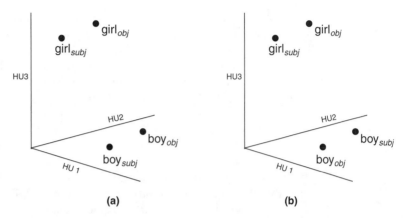

FIGURE 12.4 Hypothetical activations of hidden units for the words "boy" and "girl" when presented to the network in subject and object contexts

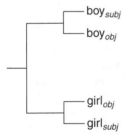

FIGURE 12.5 Cluster analysis of hidden unit activations for words "boy" and "girl" in subject and object contexts

the HU1/HU3 plane; when a word is used as the object of a sentence, its hidden unit representation lies away from that plane. This secondary distinction, which is apparent when we can directly inspect the hidden unit space, is not revealed by the hierarchical clustering. Because the "boy/girl" spatial distance is greatest, that distinction is captured by the first branching of the cluster tree. The secondary distinction, between subject and object, shows up twice—once on each branch. We have no way of knowing whether the secondary branches reflect a spatial distribution similar to the one in Figure 12.4 (a), or the one in Figure 12.4 (b). In this latter case, the instances of "boy" and "girl" as subject or object are still distinguished by spatial distance, but there does not appear to be the same *systematic use of space* to encode that distinction (notice that in the case of the "boy" patterns, the subject and object tokens are reversed, compared with the "girl" tokens). When it occurs, such systematicity is of great interest to us, and it is unfortunate that hierarchical clustering limits our ability to verify whether the the hidden unit space is being used in such a systematic way. What we really wish to do is to be able to visualize that space directly.

The problem (and the reason why hierarchical clustering is useful in the first place) is that networks we've been using for this task have 10 hidden units. Visualizing 10-dimensional space directly is difficult! Furthermore, there is no guarantee that the use of space—the dimensions along which hidden unit patterns might vary in some systematic and interesting way, lines up directly with the dimensions that are encoded by the hidden units. We might decide to settle for looking at selected 2-dimensional planes or 3-dimensional volumes in the larger dimensional space, on the assumption that any particular distinction we are interested in may require only a few spatial dimensions to encode it (so that different "cuts" through the hidden unit space might reveal different things). But there is no guarantee that the "action" in the hidden unit space aligns exactly along the axes of discrete hidden units. That is, it is possible that the state space is organized along axes which cut across the hidden unit. This is often the case with distributed representations.

This is the quandary which motivates the use of principal component analysis (PCA) of the hidden unit activations. The idea behind PCA is actually fairly straightforward.

Suppose we have hidden unit activation patterns of the sort shown in Figure 12.4. Because there are three hidden units, these patterns can be thought of as vectors (shown graphically in the figure as points) in a 3-dimensional space. Each pattern occupies a position in this 3-D space which we get by using each of the 3 hidden unit activation patterns as coordinates.

We have already pointed out that this 3-D space has been partitioned (as a result of learning) in a way which manifests two useful distinctions: (1) the difference between tokens of "boy" and "girl" and

(2) the usage of these tokens as subject or object. In this example, we appear to have three physical dimensions within which we are actually only representing two sorts of distinctions.

It is, in principle, possible that the "boy/girl" distinction and the subject/object distinction might have been captured by values along (for example), hidden units 1 and 2 (leaving hidden unit 3 unused). All values of hidden unit 1 less than some criterial magnitude (e.g., 0.5) might indicate a pattern corresponded to a "boy" token, and all values of hidden unit 2 less than some criterial magnitude might indicate the token (whether "boy" or "girl") were used as a subject. But in practice, this often does not occur. What is found more typically is the state of affairs shown in Figure 12.4(a). Although the "boy/girl" and subject/object distinctions are clearly visible in the 3-D space, the distinctions are not aligned neatly with values along any single hidden unit. The criterial values which distinguish "boy" from "girl," for example, may involve looking at several hidden units together.

On the other hand, it really does seem to be the case in this example that there are two spatial dimensions of interest. These are shown in Figure 12.6 (the hidden unit axes are shown in gray).

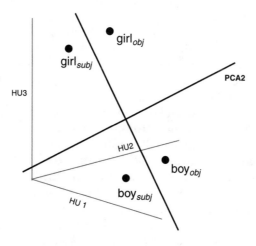

FIGURE 12.6   Two principal components for the boy/girl and subject/object distinctions

The two axes shown here represent the dimensions along which interesting variation is occurring. One dimension encodes the distinction between "boy" and "girl" and the other between subject and object. If we knew where these two axes are (we could define them in terms of our original hidden unit activation space), then we could locate each of the four points using these two new axes as a coordinate system. This new coordinate system is useful because it localizes where the information is in the hidden unit space; in a real sense, it reveals that underneath the distributed representations we often find when we look at networks, there may lurk a more localist representational scheme.

PCA is nothing more than a procedure for finding such a set of alternative axes. PCA enables us to analyze a set of vectors and discover a new set of axes along which we can represent those vectors, such that the axes capture where the variation in spatial distribution is taking place. PCA does not in any way change or move the vectors around; it just gives us another way of viewing them. Then we can look at various cuts (2-D or 3-D views) through this space. (There are as many principal components as there are dimensions in the original set of vectors. So if we have a network with 3 hidden units, there will be 3 principal components. PCA does something useful for us, however; it orders the principal components in terms of the decreasing amount of variation accounted for. Thus, the first principal component always reflects where the majority of the "action" is; the last accounts for the least variation.)

tlearn has a facility for doing PCA. Below we describe how PCA can be used to analyze the network's solution. We begin with the network which failed to find a good solution.

(1) Start **tlearn** and open the **ss5** project file. Go to **Testing Options...** in the **Network** menu and load in the weights file from the network which failed (recall that you saved this earlier in a file called **ss5-bad.wts**). The **Testing set** should be **test1.data**.

(2) Execute **Probe selected nodes** (in the **Network** menu). If you have set up your configuration file as instructed, this should print the hidden unit activations (i.e., nodes 1-10) on the **Output** window. The **test1.data** test file has 16 patterns: the 12 of interest to us, padded by an initial 4 which can be discarded (see explanation on page 243). Delete the text at the beginning of the **Output** window, plus the first 4 lines of hidden unit activations, and save the output in a file called **hidden.5-bad.vecs**. This file now

contains the vectors which we will subject to PCA. It will also be useful to have another file with labels that tell us what input produced each of the 12 activation patterns. The 12 inputs were **AABBAXXA-BXXA** but we may wish to use slightly more informative labels (to distinguish, for example, the initial A from a final A, and X's which follow A's from X's which follow B's, etc.). Create a file called **names** which contains the following labels, each on a line by itself: **As,Ae,Bs,Be,As,Xa1, Xa2,Ae,Bs,Xb1,Xb2,B3.** (We have already placed a file with these labels for you in your **tlearn** folder.)

(3) To run PCA, go to the **Special** menu and click on **Principal Components Analysis...** You will see a dialog box which should be filled out as shown in Figure 12.7. (To select a file name, double click in the rectangles to the right of **Vector file:** and **Names file:**). Then click on **Execute**.

tlearn will display the results of the PCA in a new window; the default will be a 2-D display with no labels shown, so click on **3D** and

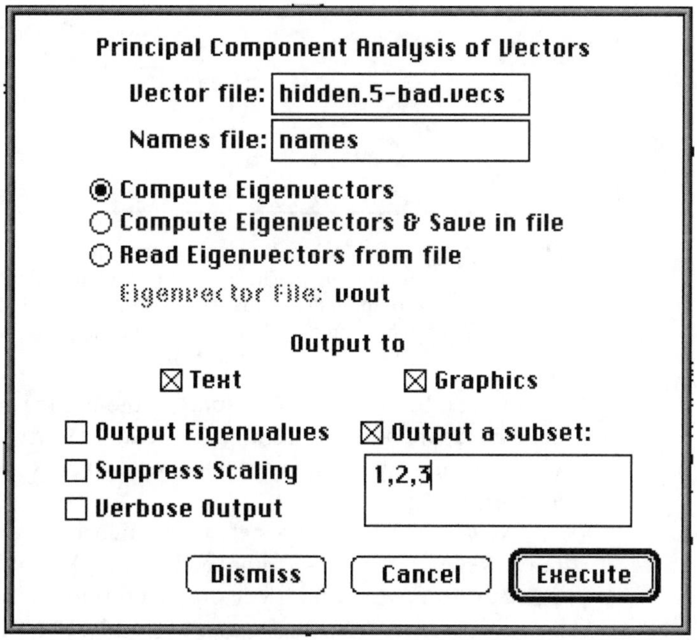

FIGURE 12.7 Principle Components Analysis dialogue box

Labels to see these. You can then rotate the 3-D image by placing the mouse within the display area, clicking and holding the button down, and moving the mouse around. You should see something similar to what is shown in Figure 12.8. (If you use our version of **hidden.5-bad.vecs** you can try to exactly recreate the perspective shown in the figure.)

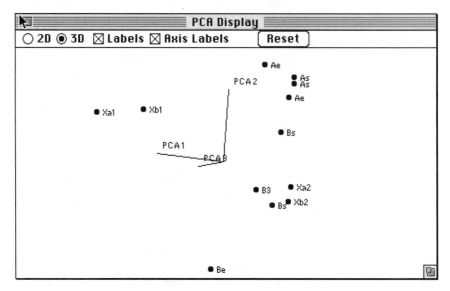

FIGURE 12.8   3-dimensional view in hidden unit activation space of a network which failed. The axes show the view along the first three principal components of this space. Points are labeled to indicate the input which produced that state. **Xa1** means the first **x** following an initial **A**; **Ae** means the final **A** in a string.

As can be seen in this display, although the **A**s and **B**s are reasonably far apart, the different types of **X**s are mixed up so that the network has no way of knowing whether a given **X** has been proceeded by an **A** or a **B**.

Now let us look at a network which has been trained incrementally. Repeat steps (1–3) above. In step (1), load in the weights file **ss.200000.wts** which you created through incremental training (or you can use our file, called **ss.5-good.200000.wts**). In step (2), save the output vectors in a file called **hidden.5-good.vecs** and use this file when you run PCA in step (3). (In our simulation, we found that the 1st, 2nd, and 4th principal components were more informative than 1,2,3, so in the PCA dialog box we checked **Output a subset** and typed in 1,2,4).

Figure 12.9 shows how the network state space has been parti-

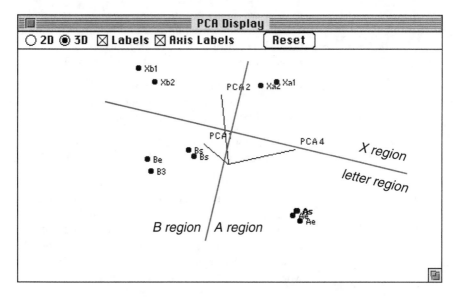

FIGURE 12.9   3-dimensional view in hidden unit activation space of a network which has been trained using the starting small technique.

tioned in a way which enables the network to keep track of the kinds of strings it is processing. The lower right portion of space has been reserved for strings which begin with **A**, and the middle right for **B** strings. When an **X** is input, the network moves to the upper region of state space; exactly which region depends on whether an **A** or **B** preceded. During the initial phase of learning, in which there were no **X**s, the network developed the **A/B** partition. This set of representations was later refined to allow for embeddings by a further partitioning.

Answers to exercises

Exercise 12.1

1. The network can recognize 3 possible inputs—**A**, **B**, and **X**—but strings can start with only **A** or **B**. These are chosen randomly, with 50/50 probability. Since the network can't know for sure what the first letter will be, its best performance will be achieved by having the outputs for both **A** and **B** units be roughly equal, and close to 0.5.

2. After the first letter has been input, the network should have learned that either the next letter will be identical to the first (in which case the string is complete), or else there may be an **X**.

3. If the network receives an **X**, there are two possible next letters: Either another **X**, or else the string may be completed by ending with the same letter that it began with (**A** or **B**, as the case may be).

4. If we've gotten to the end of the string, then we are about to get the first letter of a new string; thus the response should be as described in answer 1, above: The network should predict **A** and **B** with equiprobability. (Of course, if we found a network which *always* predicted either A or B with equiprobability, regardless of the input, we would conclude that this network had not learned at all.) This can happen.

5. On the network we trained, this is the response to the last 12 inputs:

Input	A node	B node	X node
A	0.366	0.498	0.109
A	0.473	0.179	0.305
B	0.289	0.532	0.100
B	0.119	0.521	0.382
A	0.373	0.530	0.090
X	0.280	0.055	0.874
X	0.317	0.453	0.134
A	0.413	0.292	0.221
B	0.216	0.520	0.172
X	0.184	0.144	0.800
X	0.237	0.424	0.245
A	0.224	0.550	0.153

The network is doing a miserable job. It certainly does not predict that the first letter of any string could be either an **A** or **B**.

6. It fails even on the short strings (with no intervening **X**s) to predict that the second letter will be like the first. It's not clear what this network is learning, but it's clearly not the grammar which underlies the dataset.

Exercise 12.2

*   Let's look to see how the network does at each stage during incremental learning. First, we tested the network on the **test1.data** file after it had been trained on the data with no **X**s (based on **ss.0.strings**, and using **ss.0.50000.wts**). The network's predictions to the last 12 items in the test set are shown below.

Results after training with ss.0. strings (50,000 sweeps total)

Input	A node	B node	X node
A	0.962	0.038	0.007
A	0.411	0.588	0.016
B	0.050	0.950	0.008
B	0.438	0.562	0.015
A	0.959	0.042	0.007
X	0.035	0.965	0.007
X	0.946	0.055	0.008
A	0.516	0.484	0.016
B	0.036	0.965	0.007
X	0.948	0.052	0.007
X	0.046	0.954	0.008
B	0.444	0.556	0.015

The first input is an **A**, and the next work predicts that an **A** will follow. When it receives a **B** (the third line), it predicts that a **B** will follow. In both cases, the network happened to predict correctly. When the network encounters an X, however, it becomes confused. After getting **A** and then **X**, for example, the network first predicts that the next letter will be a **B**. Then when a second **X** comes in, the network changes its mind and predicts an **A**. Performance is similarly wrong for **BXXB**.

We should not be too harsh on the network, however. After all, bear in mind that it has not yet seen any strings containing **X**s (in training). So it is hardly surprising that it should not know how the appearance of this new element interacts with the grammar it has learned from the training data.

One other thing the network does seem to have picked up is that at the end of a sequence (e.g., after the second **A**), the first letter of the next sequence can be either **A** or **B**; thus, both nodes are activated approximately equally.

- Next we did the same thing for the network after it had seen the data from **ss.1.strings** (using **ss.1.100000.wts**):

Results after training with ss.1 strings (100,000 sweeps total)

Input	A node	B node	X node
A	0.833	0.096	0.040
A	0.531	0.448	0.059
B	0.075	0.869	0.036
B	0.435	0.544	0.064
A	0.831	0.097	0.040
X	0.493	0.511	0.056
X	0.498	0.502	0.058
A	0.803	0.122	0.042
B	0.052	0.902	0.033
X	0.507	0.500	0.068
X	0.492	0.504	0.054
B	0.100	0.838	0.040

This dataset contains some occurrences (a very small percent of the time) of strings which have embedded **X**s. However, these data are apparently not sufficient to help the network learn the underlying grammar. Performance at this point is not appreciably better than with the training data that entirely lack **X**s.

- And again, the results after training with data from **ss.3.strings** (using **ss.3.150000.wts**) are shown below.

This dataset contains a larger percentage of strings with one or more **X**s embedded in them, and for the first time we begin to see some evidence that the network is able to remember the initial letter even after it has received an intervening **X**. Thus, after the sixth input (the **X** following an **A**), the network more highly activates the **A** node (0.573) than any of the other two. This is not entirely correct, since the network fails to discriminate between the likelihood of occurrence of **B** vs. **X** (**X** is possi-

Results after training with ss.3. strings (150,000 sweeps total)

Input	A node	B node	X node
A	0.747	0.097	0.171
A	0.495	0.434	0.158
B	0.112	0.730	0.188
B	0.352	0.567	0.164
A	0.745	0.100	0.170
X	0.573	0.203	0.236
X	0.253	0.466	0.246
A	0.535	0.310	0.161
B	0.089	0.730	0.203
X	0.139	0.646	0.243
X	0.361	0.332	0.249
B	0.320	0.558	0.169

ble here, **B** is not). And when a second **X** is input, the network gets confused. However, performance appears to be headed in the right direction.

- Finally, we test the network after it has completed training with data from **ss.5.strings** (using **ss.5.200000.wts**):

Results after training with ss.5. strings (200,000 sweeps total)

Input	A node	B node	X node
A	0.535	0.038	0.468
A	0.477	0.561	0.082
B	0.038	0.500	0.541
B	0.475	0.500	0.101
A	0.535	0.041	0.454
X	0.450	0.074	0.567
X	0.301	0.194	0.496
A	0.457	0.503	0.097
B	0.033	0.507	0.554
X	0.071	0.534	0.522
X	0.142	0.366	0.499
B	0.466	0.515	0.092

The network's performance is now clearly in the right ballpark. After an initial **A**, it predicts that either an **X** or an **A** might occur. If an **X** is presented, it predicts either another **X**, or the **A** that would terminate the string. And it shows the same pattern for strings beginning with B.

The fact that the activations of the correct predictions decrease after additional **X**s (e.g., after **A** and then **X**, the **A** and X node activations are 0.450 and 0.567, respectively; after a second **X** they are 0.301 and 0.496) raises an interesting question. We are tempted to infer that the memory is not yet as well-developed as it should be, and might eventually, with further training. In fact, this is probably the correct interpretation. (We could verify this by seeing how the outputs are affected by further training.)

There is another factor to be aware of, however, which might impact the change in node activations as the string length increases. In the final version of the FSA we used (i.e., the one which generated ss.5.strings), the probability of staying in the same state and emitting an **X** was 50%. But the probability of, for example, 3 successive **X**'s is less than that: 0.50 x 0.50 x 0.50, or 0.125. The probability that there will be an **X**, given that some number of **X**s precedes, thus decreases dramatically with additional **X**s. The network's goal of optimizing prediction performance can best be achieved by approximating the conditional probabilities of successor letters. The FSA that generates the grammar is a first-order Markov model which is ignorant of any prior history; and so the 0.50 probability of producing an **X** is an accurate depiction of its knowledge. In other words, the conditions attaching to the probabilities only go back one step in time. But even though the FSA that is used to generate strings has this characteristic, the actual statistical properties of those dataset so generated will also turn out to have the characteristic we just described (i.e., there is a higher likelihood of short sequences of **X**s than long sequences of **X**s). The network's sensitivity to the prior conditions or states is not restricted to the immediately succeeding item. It is thus able to compute conditional probabilities which go back further in time; in so doing, it will improve its performance. We might therefore expect that with increasing training, the node activations predicting **X** would decrease, the more **X**s in a row are input, while the node activations predicting the **A** (or **B**, as appropriate) would increase. (It's almost as if the network were getting impatient for the string to be finished!)

Installation procedures

This appendix describes the contents of the Macintosh and Windows 95 **tlearn** distribution disk and how to install **tlearn**. **tlearn** is a connectionist modelling simulation package that provides a graphical user interface (GUI) to backpropagation neural networks. There are versions of **tlearn** for Macintosh and Windows 95 and X Windows/ Unix.

Mactintosh installation

The Macintosh distribution disk contains the Macintosh **tlearn** executable and associated exercise files. The **tlearn** executable file is a FAT executable which means that it can be run on both PowerPC and 680x0 Macintosh machines. The disk also provides a set of Chapter folders which contain the project files corresponding to the exercises in this book.

Installation of **tlearn** is merely a matter of copying the contents of this disk to a new folder (e.g., a **tlearn** folder). The executable is called **tlearn** and can be launched without any other changes to the machine. Note that for simulation of large networks, it may be necessary to increase the memory usage of the application through the **Get Info...** window available within the Finder.

Windows 95 installation

The Windows 95 distribution disk contains the Windows 95 **tlearn** executable and associated exercise files. Installation of **tlearn** is merely a matter of copying the contents of this disk to a new folder (e.g., a **tlearn** folder). This is easily done with Windows Explorer. The executable is called **tlearn** and can be launched without any other changes to the machine. The disk also provides a set of Chapter folders which contain the project files corresponding to exercises in this book. We do not recommend that you use **tlearn** running under Windows 3.x (even with Win32s extensions installed).

Updates and bug reports

We do not guarantee that the software provided with this book is free of bugs. In fact, we guarantee that if you try hard enough you will find some situations where **tlearn** will break! Please tell us about any software problems you experience by emailing us at
`innate@crl.ucsd.edu`.
We may not respond to your queries immediately but we will fix the bugs as time permits. You can obtain the latest versions of **tlearn** via anonymous ftp at `crl.ucsd.edu` (in the `pub` directory) or on the world wide web at `http://crl.ucsd.edu/innate`. This site is situated in San Diego. Users from the Old World may find it easier to obtain the latest versions of **tlearn** via anonymous ftp at `ftp.psych.ox.ac.uk` (also in the `pub` directory). This site is situated in Oxford, UK.

User manual

This reference manual provides documentation on **tlearn** software version 1.0. The first section provides a quick introduction to the software and its functionality. The second section gives complete descriptions of the application's menus and dialog boxes. The third section offers a quick reference to command keys and other shortcuts. The final section offers some advice on troubleshooting and other error messages.

Introduction

tlearn is a neural network simulator which implements the backpropagation learning rule, including simple recurrent network learning and backpropagation through time, and provides a number of displays which show the network internals. **tlearn** includes a fully functional text editor as well as a number of data analysis utilities.

Configuration files

tlearn organizes files and simulations into projects. The project file is a machine readable file which stores information about option settings for training and testing. The name of the project file specifies the name of the project and acts as the prefix for other associated files for the project. There are three necessary associated files for every project. Namely the **<fileroot>.cf**, **<fileroot>.data** and

`<fileroot>.teach` files, where `<fileroot>` is the name of the project file. A complete description of these files is given in the section on *Network Configuration and Training Files*.

Editor functions

The **tlearn** text editor includes standard text editor features (e.g., Find and Replace, Cut, Copy and Paste, Revert, Go To Line, and a current line number indicator). As well as the standard features **tlearn** provides two text utilities namely **Sort...** and **Translate...** which are found in the Edit menu. Sort provides a mechanism for sorting of files with arrays of numeric values. **Translate** allows a set of find and replace actions to be carried out in an automated way. These utilities are fully described in the section *Menu and dialogue reference* where the description of their associated dialogue boxes is given. Editor short cut keys are given in the section *Command key and shortcut reference*.

Network training and testing functions

tlearn provides two types of network run modes: training and testing. These functions are found in the **Network** menu. The **Train the network** action begins training from an initial set of random weights and trains the network for the specified number of sweeps. The number of training sweeps is set in the **Training Options** dialogue box. The **Resume training** action resumes training from the current set of weights for a further specified number of sweeps. Further options and settings for training are described in the *Menu and dialogue reference* section where the **Training Options** dialogue is described.

The **Verify network has learned** action presents the trained network with a testing set specified in the **Testing Options** dialogue box. The values of the output node/s for each data presentation are given in the **Output** window. The **Probe selected nodes** action similarly presents the specified data set to the network and in this case outputs the values of the selected nodes specified in the `.cf` file. All the training and testing actions can be aborted by choosing **Abort** from the **Network** menu or the **tlearn** status display.

Network utilities

There are two Network utilities located in the Special menu. These are **Output Translation...** and **Lesioning...** . The **Output Translation...** utility allows the translation of 0/1 vectors to text. When the network operation is verified, or the network activation display is used to display network outputs, the output translation can also be given. The **Output Translation...** utility is described fully in *Menu and dialogue reference* section where the Output Translation dialogue is described.

The **Lesioning...** utility allows weight files for a specific network to be modified so that the effects of lesioning the network, i.e., removing some connections or nodes, can be examined. The **Lesioning...** utility is described fully in the *Menu and dialogue reference* section where the **Lesioning** dialogue is described.

Data analysis functions

There are two Data Analysis utilities located in the **Special** menu, namely **Cluster Analysis...** and **Principal Components Analysis....** The **Cluster Analysis...** utility allows a hierarchical clustering to be performed on a set of data vectors. The **Cluster Analysis...** utility is described fully in the *Menu and dialogue reference* section where the Cluster Analysis dialogue is described.

The **Principal Components Analysis...** utility allows the principal components of a set of data vectors to be determined and the projection of such data onto its principal components to be obtained. The **Principal Components Analysis...** utility is described fully in the *Menu and dialogue reference* section when the Principal Components Analysis dialogue is described.

tlearn displays

tlearn has seven displays which illustrate different aspects of the network simulation or show the results of some of the data analysis utilities. Three of the displays can only be selected when there is a project open as they rely on the definition of the network provided by the object files. The project dependent displays are the **Node Activation** display, the **Connection Weights** diagram and the **Network**

Architecture display. All of the displays (apart from the **Status** display) can be copied to the clipboard and pasted into word processors or other editors that can handle picture graphics.

The **tlearn Status** display (shown in Figure B.1) indicates the

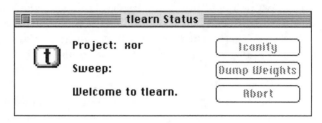

FIGURE B.1 **tlearn Status** display

name of the open project, if there is one, and the current state of **tlearn** training. The **Status** display also provides buttons to allow the display to be iconified; for the current network weights to be saved to a file; and, for the network training to be aborted

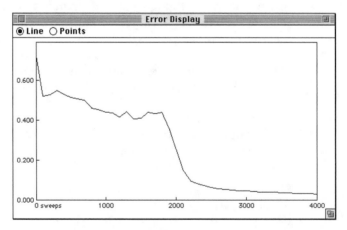

FIGURE B.2 **Error Display**

The **Error Display** (Figure B.2) gives a graph of error values for the training or testing actions carried out on the current network. Selection radio buttons at the top of the display allow the error graph to be shown as lines or as points.

The **Node Activations Display** (Figure B.3) presents the acti-

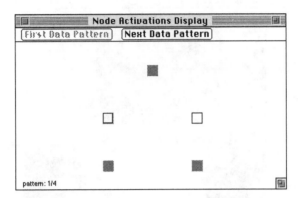

FIGURE B.3 Node Activations Display

vations of the nodes of the network as Hinton diagram displays. The orientation of the nodes on this display is specified by the network orientation set on the **Network Architecture** display. Two buttons (**First Data Pattern**, **Next Data Pattern**) allow the user to step through the training or testing patterns presented to the network.

The **Connection Weights** display (Figure B.4) displays the current network weights with a Hinton diagram showing the weight magnitudes and signs. Controls at the top of the display allow the user to specify whether the display should be updated during training, and the regularity of the diagram update, specifically updates can be set for every 1, 5, or 10 sweeps or epochs.

The **Network Architecture** display (Figure B.5) draws the network nodes and connections. The check boxes and radio buttons at the top of the display allow the following settings: The **Slabs** check box, when checked cause the input, hidden and output units to be shown as single slabs rather than as individual units. The **Labels** check box toggles the display of node labels. The **Arrows** check box toggles the appearance of arrow heads on the connection lines. The **Bias** check box toggles the appearance of the bias node. The **Orient:** radio buttons specify the orientation of the network display as bottom-to-top or left-to-right. The orientation specified here is also used in the **Node Activations** display.

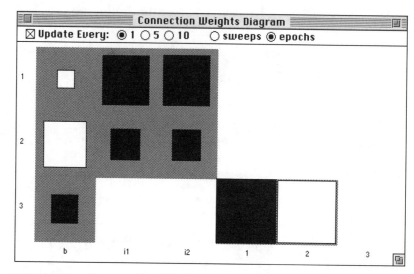

FIGURE B.4   Connection Weights Diagram

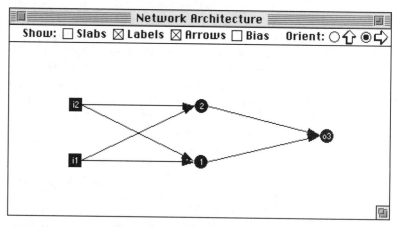

FIGURE B.5   Network Architecture display

The **Cluster Display** (Figure B.6) is used to present the output of the **Cluster Analysis...** utility (found in the **Special** menu). **Cluster analysis** generates a cluster diagram which is a tree illustrating the clusters of the data. The **Distances** check box toggles the appearance of distance values on the cluster diagram in the display.

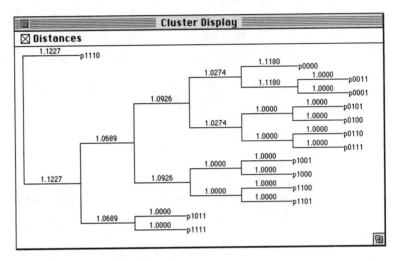

FIGURE B.6   Cluster Display

The **Cluster Analysis...** utility is described more fully in the *Menu and dialogue reference* section, where the **Cluster Analysis** dialogue is described.

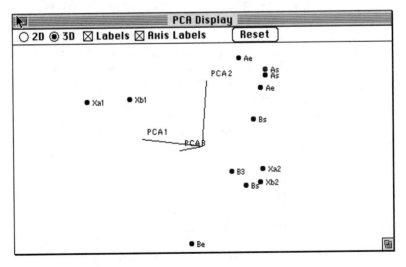

FIGURE B.7   PCA Display

The **PCA Display** (Figure B.7) is used to present the output of the **Principal Components Analysis...** utility (found in the **Special** menu). The output of principal components analysis is the projection of the data points onto their principal components. The **2D/3D** radio buttons allow switching between 2 and 3 dimensioned projections. The **Labels** check box toggles the appearance of labels for the data points. The **Axis Labels** check box toggles the appearance of labels for the axes shown in the display. The **Reset** button is used for **3D** projections to return the projection to the standard orientation. For **3D** projections the orientation of the axes can be modified by click-dragging on the diagram which rotates the projection accordingly. The **Principal Components Analysis...** utility is described more fully in *Menu and dialogue reference* section, where the **PCA** dialogue is described.

Network configuration and training files

tlearn requires three input files: the network configuration—the **cf** file, the input pattern—the **data** file, the output (teacher) pattern—the **teach** file and the project file. An additional input file may be used to specify a reset schedule for context nodes—the **reset** file. **tlearn** may also create output files for weights, error, node activations, etc. All files should begin with the same name; this is referred to as the *fileroot*. The project file is called **<fileroot>**. The project file is created automatically when the user starts a new project in **tlearn**. The different files are distinguished by having different extensions (where an extension consists of a period followed by a fixed designator). Every **tlearn** simulation will contain at least the following 4 files:

```
<fileroot>
<fileroot>.cf
<fileroot>.data
<fileroot>.teach
```

The optional reset file is called:

```
<fileroot>.reset
```

Network configuration (`.cf`) file

This file describes the configuration of the network. It must conform to a fairly rigid format, but in return offers considerable flexibility in architecture. There are three sections to this file. Each section begins with the keyword in upper case, flush-left. The three section keywords are **NODES:**, **CONNECTIONS:**, and **SPECIAL:**. Note the colon. Sections must be described in the above order.

NODES:

This is the first line in the file. The second line specifies the total number of nodes in the network as "**nodes = #**". Inputs do *not* count as nodes. The total number of inputs is specified in the third line as "**inputs = #**". The fourth line specifies the number of nodes which are designated as outputs according to "**outputs = #**'. (Note that these two lines essentially give the lengths of the **.data** and **.teach** vectors.) Lastly, the output nodes are listed specifically by number (counting the first node in the network as 1) in the order that the **.teach** information is to be matched up with them. The form of the specification is "**output nodes are <node-list>**". (If only a single output is present one can say "**output node is #**"). If no output nodes are present, this line is omitted. Spaces are critical.

  Node number can be important for networks in which there are fixed copy-back links. Copy-back links allow for saving of node activations so that they can be used on the next sweep. Because node activations are calculated in ascending order, with the order determined by the number of the node, it is important that node activations be saved *after* they are calculated. It is also important that a unit which receives input from a node which is serving as a state/context node (and has thus storing some other nodes activation from the previous time cycle) calculate its activation before the state/context node gets updated on the current sweep. Both considerations lead to the following rule of thumb: Any node receiving input from a state/context node must have a *lower* node number than the state/context node. This is illustrated in the example .cf file at the end of this section.

CONNECTIONS:

This is the first line of the next section. The line following this must specify the number of groups, as in "**groups = #**" (All connections in a group are constrained to be of identical strength; e.g., as in the translation invariance problem in Chapter 7; in most cases **groups = 0**.) Following this, information about connections is given in the form:

`<node-list> from <node-list> [= <fixed> | <group #> | <min & max>]`

If values are specified for **<min & max>** (e.g., -5.0 & 5.0) then the weights for the relevant connections will not be allowed to exceed these minimum and maximum values. Weights specified as **fixed** will have values fixed at their initialization values (if **<min & max>** are set to 1.0 & 1.0, then the weights are set to 1.0 and remain unchanged throughout learning; this is typically used for connections from context units).

It is also possible to say:

`<node-list> from <node-list> = <min> & <max> fixed one-to-one`

This last form is used, e.g., when node 1 is fed from node 4, node 2 is fed from node 5, and node 3 is fed from node 6, as opposed to the usual case of node 1 being fed from nodes 4–6, node 2 being fed by nodes 4-6, and node 3 being fed by nodes 4–6.

A **<node-list>** is a comma-separated list of node numbers, with dashes indicating that intermediate node numbers are included. A **<node-list>** contains *no spaces*. Nodes are numbered counting from 1. Inputs are likewise numbered counting from 1, but are designated as "i1", "i2", etc. Node 0 always outputs a 1 and serves as the bias node. If biases are desired, connections *must* be specified from node 0 to specific other nodes (not all nodes need be biased). Groups must be labeled by integers ascending in sequence from 1. It is also permissible to say

`<group #> = <min & max>`

provided that the group has already been completely defined.

SPECIAL:

This is the first line of the third and final section. Optional lines can be used to specify whether some nodes are to be linear ("**linear = <node-list>**"), which nodes are to be bipolar ("**bipolar = <node-list>**")[1], which nodes are selected for special printout ("**selected = <node-list>**"), and the initial weight limit on the random initialization of weights ("**weight_limit = <#>**"). Again, *spaces are critical*.

Example **.cf** files are given at the end of this section for several network architectures.

Network data (.data) file

This file defines the input patterns which are presented to **tlearn**. The first line must either be "**distributed**" (the normal case) or "**localist**" (when only a few of many input lines are nonzero). The next line is an integer specifying the number of input vectors to follow. Since exactly one input vector is used for each time step, this is equivalent to specifying the number of time steps. The remainder of the **.data** file consists of the input. These may be input as integers or floating-point numbers.

In the (normal) "**distributed**" case, the input is a set of vectors. Each vector contains n_i floating point numbers, where n_i is the number of inputs to the network. Note that these input vectors are always used in the exact order that they appear in the **.data** file (unless the randomization option is specified).

In the "**localist**" case, the input is a set of **<node-list>**s (defined below) listing only the numbers of those nodes whose values are to be set to one. Node lists follow the conventions described in the **.cf** file. An example **.data** file is shown below in both the "**localist**" and "**distributed**" case.

---

1. Linear nodes simply output the inner-product of the input and weight vectors, or *net*. Logistic units are sigmoidal: The activation function for each node is $y = 1/(1 + e^{-net})$. Logistic node output is bounded by 0 and 1. Bipolar nodes have an extended range—their output ranges continuously from -1 to +1. The activation function for bipolar nodes is $y = (2/(1 + e^{-net})) - 1$.

```
distributed
4
1 0 0 0 0
0 1 1 0 0
0 1 0 0 0
1 1 1 0 1

localist
4
1
2,3
2
1-3,5
```

Network teach (.teach) file

This file is required whenever learning is to be performed. As with the **.data** file, the first line must be either "**distributed**" (the normal case) or "**localist**" (when only a few of many target values are nonzero). The next line is an integer specifying the number of output vectors to follow.

In the (normal) "**distributed**" case, each output vector contains n_o floating point numbers, where n_o is the number of outputs in the network. An asterisk ("*****") may be used in place of a floating point number to indicate a "don't care" output. In patterns containing "don't care" indicators, no error will be generated for those output units for which a ***** is specified.

In the "**localist**" case, each output vector is a set of **<node-list>**s whose targets are a 1 as opposed to a 0. Node lists follow the conventions described in the **.cf** file. An example **.teach** file for a network with one output unit is given below:

```
distributed
4
0.1
0.9
*
0.
```

Network reset (.reset) file

This file is required whenever the context nodes need to be reset to zero. As with the **.teach** file, the first line must be an integer specifying the number of time stamps to follow. Each time stamp is an integer specifying the time step (i.e., pattern number) at which the network is to be completely reset. As with the **.teach** file, the time stamps must appear in ascending order. An example **.reset** file with 2 time stamps (patterns 0 and 3) is given below:

```
2
0
3
```

Weights (<fileroot>.<runs>.wts) file

At the conclusion of a **tlearn** session, the results of training are saved in a "weights file." This file name incorporates the fileroot, the number of learning sweeps (runs) which resulted in this network, and ends with "**wts**" as the literal extension. This file contains weights and biases resulting from training. Weights are stored for every node (except the bias node, 0), from every node (including the bias node). A sample weights file for an XOR (**2x2x1**) network is shown below. (Sources are shown explictly here for connections into node 1 only; they do not appear in the actual weights file.)):

```
NETWORK CONFIGURED BY TLEARN
# weights after 10000 sweeps
# WEIGHTS
# TO NODE 1
-6.995693          (from bias node)
4.495790           (from input 1)
4.495399           (from input 2)
0.000000           (from node 1)
0.000000           (from node 2)
0.000000           (from node 3)
# TO NODE 2
2.291545
-5.970089
-5.969466
0.000000
0.000000
0.000000
# TO NODE 3
4.426321
0.000000
0.000000
-9.070239
-8.902939
0.000000
```

This file can also be produced by requesting periodic check-pointing (dumping of a weights file) either in order to recreate intermediate stages of learning, or to avoid having to re-run a lengthy simulation in the event of premature termination. This weights file can be loaded into **tlearn** in order to test with a trained network.

Error (.err) file

If error logging is requested, a file will be produced containing the RMS error, saved at user-specifiable intervals.

Example .cf files

EXAMPLE 1:

This illustrates a feed-forward network which implements a **2x2x1**
XOR network (cf. Chapter 4). Notice that in **tlearn**, the 2 inputs are
not nodes; the network itself has only 2 hidden nodes and 1 output
node. (There are still learnable connections from the 2 inputs to the 2
hidden nodes.)

```
NODES:
nodes = 3
inputs = 2
outputs = 1
output node is 3
CONNECTIONS:
groups = 0
1-3 from 0
1-2 from i1-i2
3 from 1-2
SPECIAL:
selected = 1-2
weight_limit = 1.0
```

EXAMPLE 2:

This illustrates a network that receives 3 inputs, has 4 hidden nodes, 2
output nodes, and 4 copy-back nodes; each copy-back node receives
the activation of the corresponding hidden node at the prior cycle.
Notice that the copy-back nodes are linear, receive no bias, and have
fixed downward connections from the hidden nodes. In the number
scheme, **i1-i3** designate the 3 inputs; nodes **1-4** are the hidden
nodes; nodes **5-6** are the output nodes; and nodes **7-10** are the copy-
back (state/context) nodes.

```
NODES:
nodes = 10
inputs = 3
outputs = 2
output nodes are 5-6
```

```
CONNECTIONS:
groups = 0
1-6 from 0
1-4 from i1-i3
1-4 from 7-10
5-6 from 1-4
7-10 from 1-4 = 1. & 1. fixed one-to-one
SPECIAL:
linear = 7-10
weight_limit = 1.
selected = 1-4
```

EXAMPLE 3:

This illustrates a network which receives 9 inputs, has 3 hidden nodes (**1-3**) and 1 output node (**4**). The 3 hidden nodes have limited receptive fields; each one receives connections from only 3 of the inputs. In addition, the connections are grouped (i.e., trained to assume the same values), thus benefiting from the learning that occurs for other nodes in the group (e.g., even when deprived of input). The result is that each hidden node has 3 different input weights; each of the 3 weights has a similar weight leading into the other 2 hidden nodes. This scheme is similar to the translation invariance network in Chapter 7. Finally, weights are confined to the range **-5/+5**.

```
NODES:
nodes = 4
inputs = 9
outputs = 1
output node is 4
CONNECTIONS:
groups = 3
1-4 from 0
1 from i1 = group 1
1 from i2 = group 2
1 from i3 = group 3
2 from i4 = group 1
2 from i5 = group 2
2 from i6 = group 3
```

```
3 from i7 = group 1
3 from i8 = group 2
3 from i9 = group 3
4 from 1-3
group 1 = -5 & 5
group 2 = -5 & 5
group 3 = -5 & 5
SPECIAL:
selected = 1-3
weight_limit = 0.1
```

Menu and dialogue reference

The File menu

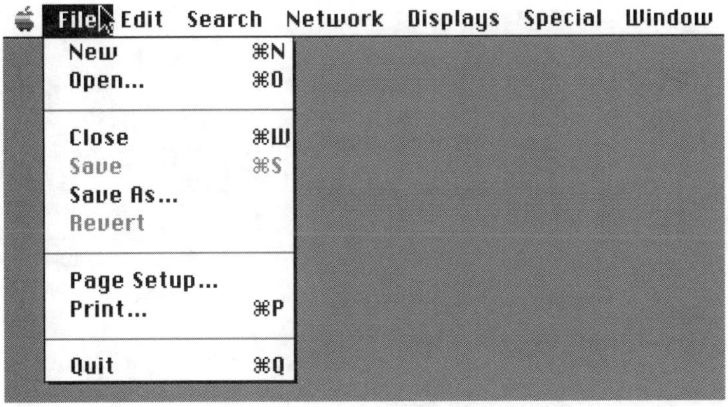

FIGURE B.8   **File** menu

Figure B.8 shows the **File** menu and associated commands which act as follows:

New   Creates a new text file window.

Open...   Brings up the **File Open** dialogue box to allow selection of text files for editing.

Close   Closes the current window, either a text window or a display.

Save	Saves the current text window.
Save As...	Brings up the **Save As** dialogue box where a name for saving the current text window can be selected.
Revert	Reverts to the previously saved version of the file.
Page Setup...	Brings up the **Page Setup** dialogue box.
Print...	Prints the current window. Text windows and displays (apart from the **Status** display) can be printed.
Quit	Quits `tlearn`.

The Edit menu

Figure B.9 shows the **Edit** menu. All but the last two commands

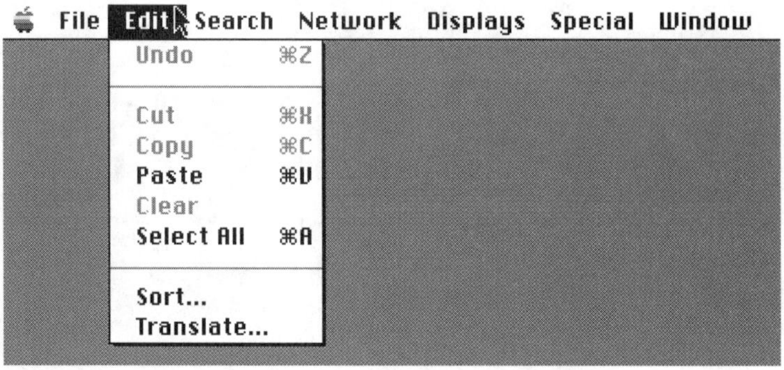

FIGURE B.9   Edit menu

(**Sort...** and **Translate...**) are standard. Note that the **Copy** command can also be used to copy displays to the clipboard so they can be pasted elsewhere (e.g.. word processors, graphics editors). **Sort...** and **Translate...** are text utilities used for manipulating text files for use with `tlearn`.

The **Sort...** command brings up the **Sort** dialogue (shown in Figure B.10) which is used to specify settings for a **sort** action to be applied to the current text window. Sorting can only be applied to text files which contain a couple array of numerical values, that is each line of the text window must contain an equal number of numerical

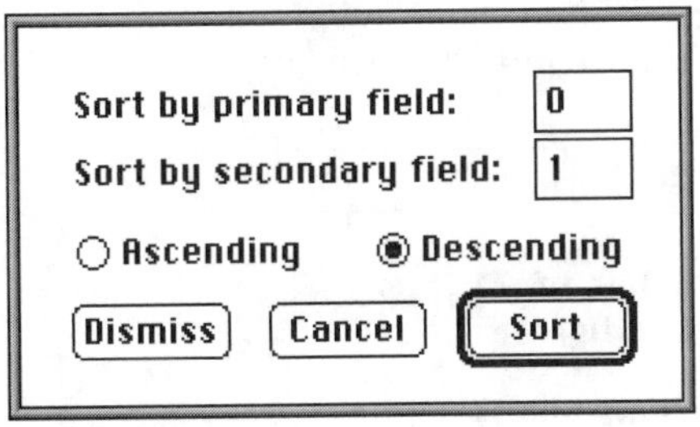

```
┌──────────────────────────────────────────┐
│ ┌────────────────────────────────────────┐ │
│ │                                        │ │
│ │  Sort by primary field:    [ 0 ]       │ │
│ │  Sort by secondary field:  [ 1 ]       │ │
│ │                                        │ │
│ │  ○ Ascending   ● Descending            │ │
│ │  [Dismiss]  [ Cancel ]  [ Sort ]       │ │
│ │                                        │ │
│ └────────────────────────────────────────┘ │
└──────────────────────────────────────────┘
```

FIGURE B.10 **Sort** dialogue box

values separated by white space. The counting of fields begins at zero (0) rather than one (1). Sorting can be done over a primary field or a primary and secondary field, and the lines can be sorted in ascending or descending order. As with many of the dialogue boxes the **Sort** dialogue settings can be specified but dismissed (by pressing the **Dismiss** button). Instead of executing the action immediately, the **sort** dialogue simply saves the settings. Pressing the **Cancel** button reverts the settings to the previous values. Pressing the **Sort** button causes the **sort** action to be done. The lines of text are sorted and the result is returned into a new text window which is called "**<WindowName> Sorted**" where **<WindowName>** is the name of the window being sorted. Note that the text of the sorted window is not saved to a file, but when the file is saved, a **Save As** dialogue prompts the user for a filename for the new text window. An example of the use of **Sort...** is given in Chapter 5.

The **Translate...** command brings up the **Translate** dialogue (shown in Figure B.11) for a translation action to be applied to the current text window. The translation requires a pattern file which contains lines which specify the translations to be performed. The format of the lines which specify the translation is as follows:

```
<find string> <replace_string>
```

The **find** string cannot contain spaces, but the **replace** string, which is all of the line apart from the first word or string, can contain spaces. An example pattern file is:

```
JE Jeff Elman
KP Kim Plunkett
TL tlearn
BP Backpropagation
```

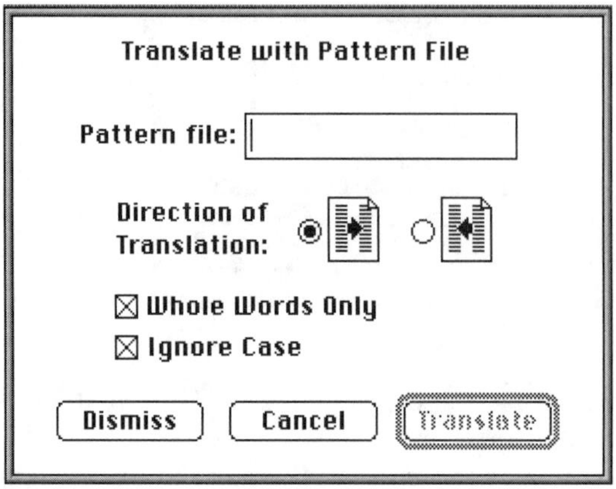

FIGURE B.11 **Translate** dialogue box

The translation action is performed in the following manner: A new window with a copy of the text of the current window is displayed (the new window is given the name of the current window with the word "Translated" appended to it), then each line of the pattern file is used in turn and the translation is applied to the new window. If the direction of translation is from left to right then occurrences of the **find** string are replaced by the **replace** string. If the direction of translation is from right to left then occurrences of the **replace** string are replaced by the **find** string. The **Whole Words Only** check box when checked ensures that the string being searched for are whole words only, that is, the string has surrounding white space. The **Ignore Case** check box allows the searching to find strings regardless of the mixture of upper and lower case letters. An example of the use of **Translate** is given in Chapter 8.

The Search menu

Figure B.12 shows the **Search** menu. The **Search** menu commands

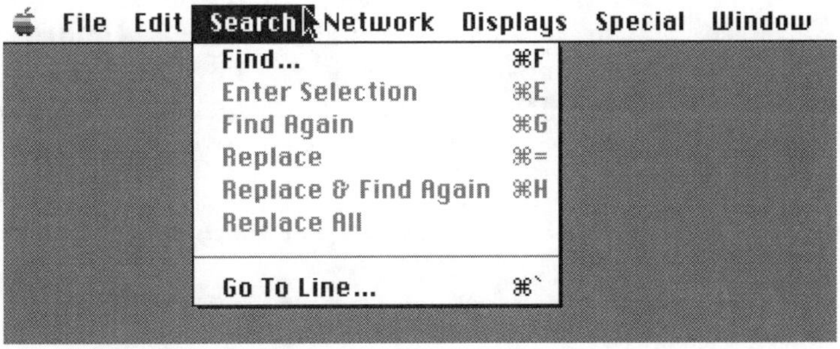

FIGURE B.12   The **Search** menu

act as follows:

Find...   Brings up the **Find and Replace** dialogue (shown in Figure B.13), where the **Find** string and the **Replace** string can be entered and conditions for document searching can be set.

Enter Selection   Copies the current selection into the **Find** string. If no text is currently selected, then this menu item is disabled.

Find Again   Repeats the previous search in the current search direction.

Replace   Replaces the currently selected **Find** string with the **Replace** string. If the current selection is empty or not equal to the **Find** string then no replacement is made. Hence, the **Replace** action is only sensibly done after a successful **Find** action.

Replace & Find Again   Replaces the selected **Find** string with the **Replace** string and searches for the next occurrence of the **Find** string.

Replace All   Replaces all occurrences of the **Find** string from the current cursor position to the end of the document. If the **Wrap Around** option is chosen then all occurrences of the **Find** string in the whole document are replaced.

Go To Line...   Brings up the **Go To Line** dialogue where the cursor is moved to the line number entered.

When a **Find** string has been entered, then the **Find** button can be pressed and a search begins for occurrences of the **Find** string in the current window starting from the current cursor position. If an occurrence of the **Find** string is found, then the search stops and the occurrence is highlighted. If no occurrence is found then the system beeps. Two of the options in the **Find** and **Replace** dialogue are the same as those found in the **Translate...** dialogue; namely **Whole Words Only**, which ensures that the occurrences of the **Find** string found have surrounding white space; and **Ignore Case**, which allows searching to find strings regardless of upper or lower case letters. The **Wrap Around** option causes searching which reaches the end of the current window to begin again from the start of the document.

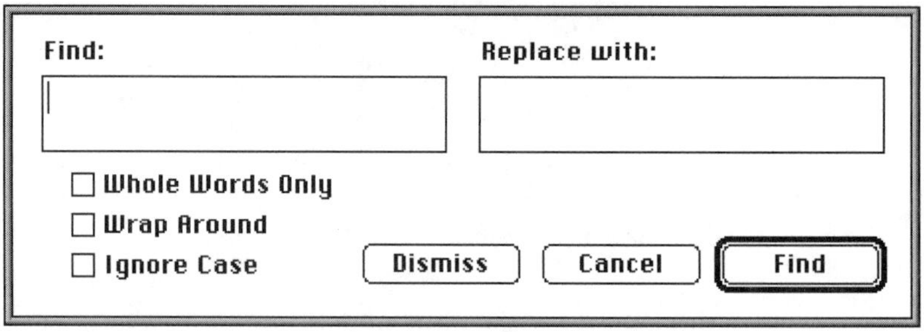

Find: Replace with:

□ Whole Words Only
□ Wrap Around
□ Ignore Case   Dismiss   Cancel   Find

FIGURE B.13   The **Find and Replace** dialogue box

The Network menu

Figure B.14 shows the **Network** menu. The **Network** menu com-

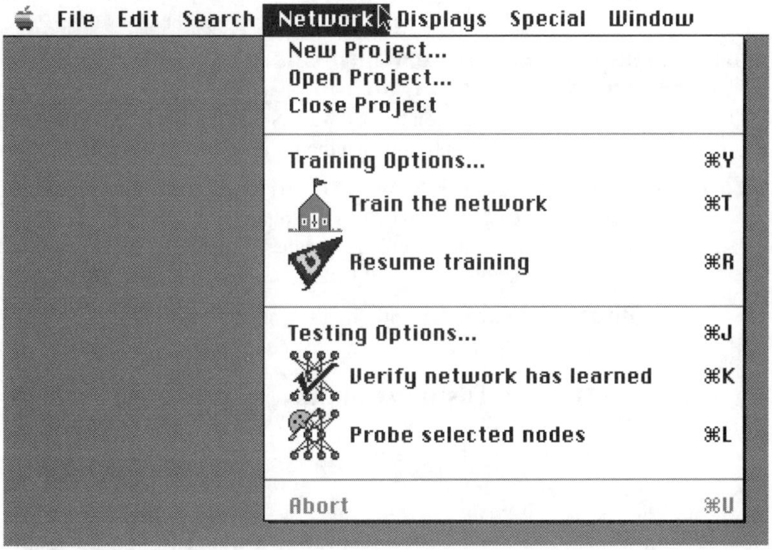

FIGURE B.14 The **Network** menu

mands act as follows:

New Project... Brings up the **New Project** dialogue box, which allows the entry of a new project name. Once the name is chosen, **tlearn** opens three new text windows associated with the project, that is, the **<name>.cf**, **<name>.data** and **<name>.teach** files. If these files already exist in the directory in which the project was opened then the files are opened, otherwise the windows are empty.

Open Project... Brings up the **Open Project** dialogue box, which allows the selection of **tlearn** project files. Once a project is chosen for opening, tlearn opens the three associated **.cf**, **.data** and **.teach** text files.

Close Project Sets the current project to none and closes the current projects **.cf**, **.data** and **.teach** file windows, if they are currently open.

Training Options... Brings up the **Training Options** dialogue box. The **Training Options** dialogue box is described in more detail below.

Train the network	Begins training the network according to the training options set in the **Training Options** dialogue box.
Resume training	Resumes training from the current set of network weights.
Testing Options...	Brings up the **Testing Options** dialogue box. The **Testing Options** dialogue box is described in more detail below.
Verify network has learned	Runs the **Verify Network** action which presents the testing set (specified in the **Testing options** dialogue) and calculates the output values of the output units and prints these values into the **Output** window. If the **Output** window is not currently opened then it is opened and selected.
Probe selected nodes	Runs the **Probe Network** action which presents the testing set (specified in the **Testing Options** dialogue) and calculates the output values of the selected nodes and prints these values into the **Output** window. The selected nodes are specified in the .cf file.
Abort	Aborts a currently running network action.

The Displays menu

Figure B.15 shows the **Displays** menu. The display menu indicates

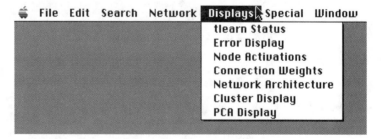

FIGURE B.15   The **Displays** menu

with check marks which displays are currently being shown by **tlearn**. Selecting an item in the **Displays** menu either hides the display, if it is currently being shown, or shows the display if it is currently hidden. Note that a display window, if it is currently shown, can be brought to the front by selecting it from the **Window** menu.

The Special menu

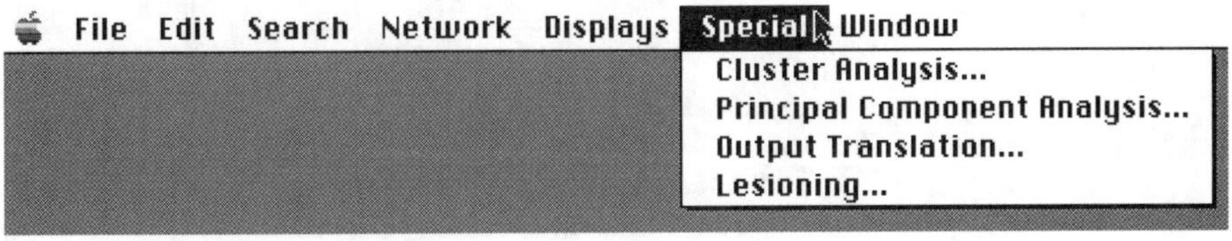

File   Edit   Search   Network   Displays   **Special**   Window

                                       Cluster Analysis...
                                       Principal Component Analysis...
                                       Output Translation...
                                       Lesioning...

FIGURE B.16   The **Special** menu

Figure B.16 shows the **Special** menu. The **Special** menu commands act as follows:

Cluster Analysis...   Brings up the **Cluster Analysis** dialogue. The **Cluster Analysis** action allows a hierarchical clustering to be performed on a set of vectors specified in a file and for the results of the clustering to be displayed in the form of a cluster diagram which shows the tree of clusters that are identified by the clustering method. The **Cluster Analysis** dialogue box is described in more detail later in this section

Principal Component Analysis...   Brings up the **Principal Component Analysis** dialogue. The **PCA** action allows the principal components of a set of vectors to be calculated and the projection of the vectors onto these principal components to be output or displayed graphically. The **Principal Components Analysis** dialogue box is described in more detail below.

Output Translation...   Brings up the **Output Translation** dialogue box. This dialogue allows the setting of the pattern file for an output translation and a setting of the **Output Mapping Criteria**. The **Output Translation** dialogue box is described below.

Lesioning...   brings up the **Lesioning** dialogue box. The **Lesioning** action allows the lesioning of a saved weight file which can selectively remove a random proportion of a set of the network nodes or connections. The **Lesioning** dialogue box is describe below.

The **Cluster Analysis...** dialogue is shown in Figure B.17. The items

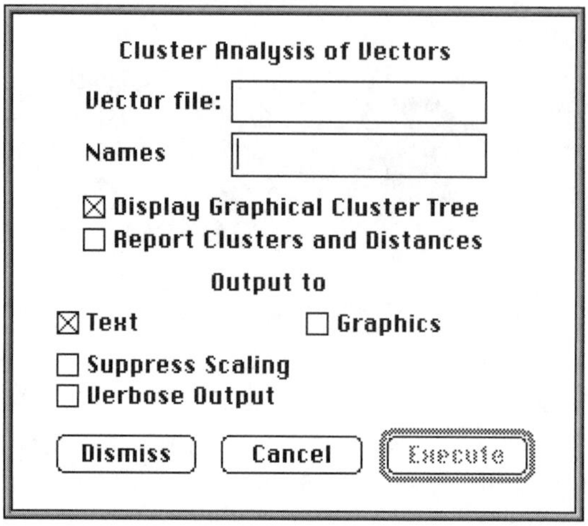

FIGURE B.17   **Cluster Analysis** dialogue box

in the dialogue are:

Vector file: Specifies the file which contains the data vectors upon which the clustering is to be done.The file name can be entered by typing the file name directly, or by double-clicking on the entry box and choosing the file name from a file selection box.

Names   Specifies a file of names to be associated with the data vectors. These names are used in the cluster diagram that is displayed. The file name can be entered by typing the file name directly, or by double-clicking on the entry box and choosing the file name from a file selection box.

Display Graphical Cluster Tree   Toggles whether the cluster tree is displayed or not.

Report Clusters and Distances   Toggles whether the clusters and distances are reported in the text output and whether the distance values are reported in the graphics output.

Output to Text/Graphics   These check boxes specify whether the cluster analysis output is sent to the **Output** text window or the **Cluster Display** or both.

Suppress Scaling   This check box suppressing scaling in the cluster analysis.

Verbose Output   This check box toggles verbose output for the text output.

Once the settings for the **Cluster Analysis** have been set, then the cluster analysis can be performed. Either the text **Output** window or the **Cluster Display** window or both is brought to the front and the cluster diagram (and other text output, if specified) is given. An example of the use of **Cluster Analysis** is given in Chapter 6.

The **Principal Component Analysis...** dialogue is shown in

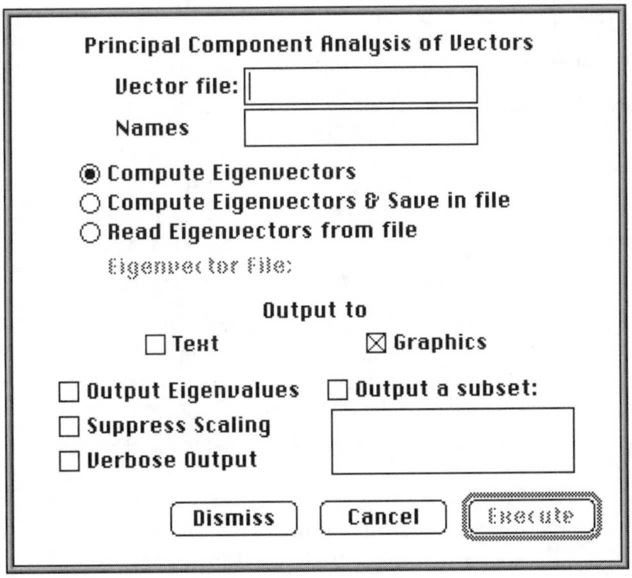

```
┌──────────────────────────────────────────────────┐
│ ╔════════════════════════════════════════════════╗ │
│ ║   Principal Component Analysis of Vectors       ║ │
│ ║                                                  ║ │
│ ║   Vector file: │                                ║ │
│ ║                                                  ║ │
│ ║   Names        ┌─────────────────────┐          ║ │
│ ║                                                  ║ │
│ ║   ⦿ Compute Eigenvectors                        ║ │
│ ║   ○ Compute Eigenvectors & Save in file         ║ │
│ ║   ○ Read Eigenvectors from file                 ║ │
│ ║       Eigenvector file:                          ║ │
│ ║                                                  ║ │
│ ║              Output to                           ║ │
│ ║       □ Text          ⊠ Graphics                ║ │
│ ║                                                  ║ │
│ ║   □ Output Eigenvalues   □ Output a subset:     ║ │
│ ║   □ Suppress Scaling                             ║ │
│ ║   □ Verbose Output                               ║ │
│ ║                                                  ║ │
│ ║     ( Dismiss )  ( Cancel )  ( Execute )         ║ │
│ ╚════════════════════════════════════════════════╝ │
└──────────────────────────────────────────────────┘
```

FIGURE B.18   Principal Component Analysis dialogue box

Figure B.18. The items in the **PCA** dialogue are:

Vector file: Specifies the file which contains the data vectors upon which the principal components analysis is to be done. The file name can be entered by typing the file name directly, or by double-clicking on the entry box and choosing the file name from a file selection box.

Names Specifies a file of names to be associated with the data vectors. These names are used in the projection of the vectors onto their principal components. The file name can be entered by typing the file name directly, or by double-clicking on the entry box and choosing the file name from a file selection box.

Compute Eigenvectors/ Compute Eigen- vectors & Save in file/Read Eigenvectors from file	These radio buttons specify whether the eigenvectors of the analysis are computed; computed and saved in a file; or, not computed but read from a supplied file.
Eigenvector file	Specifies the file where the eigenvectors are saved or read.
Output to Text/ Graphics	Specify whether the principal component analysis output is sent to the output text window, or the **PCA Display**, or both.
Output Eigenvalues	Toggles whether the eigenvalues are printed in the text output.
Suppress Scaling	Toggles whether scaling is suppressed in the principal components analysis.
Verbose Output	Toggles **Verbose Output** for the text output.
Output a subset	Toggles whether only a set of the principal components are output or displayed. For example, if the user wants to display the projection of the vectors onto their second, third and fourth principal components, then the text "**2, 3, 4**" can be entered in the text box here and the **Output a subset**: check box can be clicked so the **PCA display** will show the projection onto principal components 2, 3, and 4.

Output Translation... brings up the **Output Translation** dialogue which is shown in Figure B.19. This dialogue allows the setting of the pattern file for an **Output Translation** and a setting of the **Output Mapping Criteria**. The items in the **Output Translation** dialogue are described here.

Pattern file	Specifies the **Output Translation** definition file. The format of the **Output Translation** file is given below. An example **Output Translation** file is described in Chapter 11.
Threshold Output/ Euclidean Distance	These radio buttons specify the **Output Mapping Criteria**. In **Threshold Output** mode, when the **Output Translation** is applied, a threshold function is applied to the elements of the output vector converting the values to 0 and 1. If the converted output vector is not present in the **Output Translation** mapping then the **Output Translation** includes a question mark character. In **Euclidean Distance** mode, no function is applied to the output vector. Instead the **Output Translation** is determined by the closest vectors according to Euclidean distance in the **Output Translation** mapping.

The **Output Translation** utility allows the specification of arbitrary translations of 0/1 vectors to text. The **Output Translation** def-

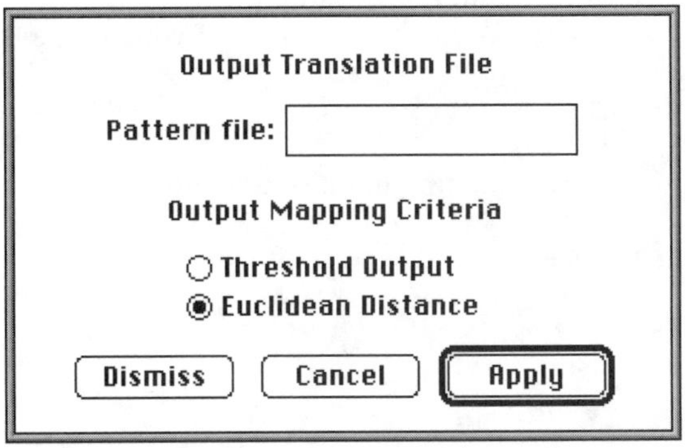

FIGURE B.19 **Output Translation** dialogue box

inition file allows the user to specify the splitting up of the output vector, permitting each part of the output vector to be assigned to different mappings, which are also specified in the file.

The format of the **Output Translation** definition file is as follows: The file begins with a **MAPPINGS:** section which specifies how the output vector is split up and by which mappings the parts of the output vector are translated. The format of lines in the **MAPPINGS:** section is as follows:

<node_list> from <MAPPING_NAME>

where **<node_list>** is a specification of a contiguous set of outputs, e.g., **1-4**; **<MAPPING_NAME>** is any unique name for a mapping which will be specified in the file.

An example **MAPPINGS:** section of an **Output Translation** definition would be:

MAPPINGS:
1-6 from PHONEMES
7-12 from PHONEMES
13-18 from PHONEMES
19-20 from SUFFIXES

Here a mapping **PHONEMES** is to be used for nodes 1 to 6, 7 to 12 and 13 to 18 and a mapping **SUFFIXES** is to be used for nodes 19 to 20.

Following the **MAPPINGS:** section come each of the mappings for the **Output Translation**. Each mapping begins with a line which contains its name followed by a colon ("`:`") then any number of lines to specify the vectors to be mapped which take the following format:

```
<Label> <Vector>
```

An example mapping definition would be:

```
SUFFIXES:
W 0 0
X 1 0
Y 0 1
Z 1 1
```

The **Lesioning** dialogue box is shown in Figure B.20. The items in the **Lesioning** dialogue are:

Weight File: Specifies the weight file on which the lesioning is to be performed.

NODES: Specifies whether any nodes are to be lesioned. The **Location:** entry box is used to specify the nodes that are to be lesioned. Any list of node numbers can be entered in the format of node lists that is used in the `.cf` file. If no list of nodes is given then all the nodes are assumed to be chosen. The **% removal** entry box specifies the proportion of the nodes from the node list that are to be randomly chosen to be lesioned.

CONNECTIONS: Specifies whether any connections are to be lesioned. The **Location:** entry box is used to specify the connections that are to be lesioned. A comma separated list of connections specifications in the format:

```
<node-list> from <node-list>
```

as used in the **CONNECTIONS:** section of a `.cf` file can be entered here. If no list of connections is given then all connections are assumed to be chosen. The **% removal** entry box specifies the proportion of the connections that are to be randomly chosen to be lesioned.

When the **Lesioning** settings have been completed and the **Lesion** button is pressed then the weights from the specified weights file are read in. If the **NODES:** check box was set, then the nodes to be lesioned are randomly chosen and the chosen nodes are removed from the network. This means that the nodes to be lesioned have all connections (both input and output connections) set to zero. If the **CONNECTIONS:** check box was set, then the connections to be lesioned are randomly chosen and the chosen connections are set to zero. The

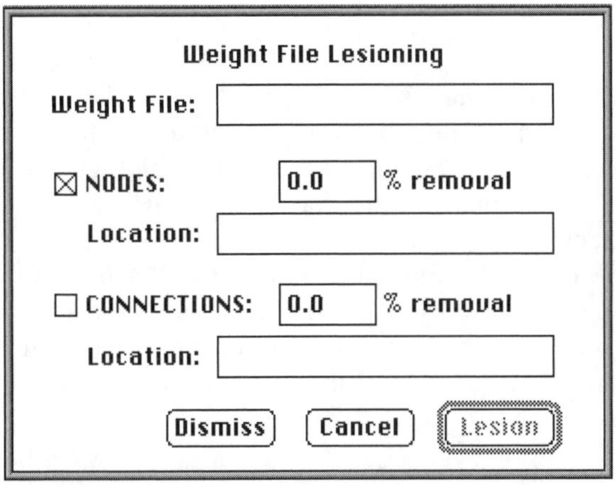

```
┌─────────────────────────────────────────────┐
│ ┌─────────────────────────────────────────┐ │
│ │           Weight File Lesioning           │ │
│ │ Weight File: [                          ] │ │
│ │                                           │ │
│ │ ⊠ NODES:      [0.0  ] % removal            │ │
│ │   Location:  [                          ] │ │
│ │                                           │ │
│ │ ☐ CONNECTIONS: [0.0  ] % removal           │ │
│ │   Location:  [                          ] │ │
│ │                                           │ │
│ │      (Dismiss)  (Cancel)  (Lesion)        │ │
│ └─────────────────────────────────────────┘ │
└─────────────────────────────────────────────┘
```

FIGURE B.20 **Lesioning** dialogue box

newly lesioned set of weights is displayed in a new window entitled **<name>.lesion.wts**. This window is an unsaved text window which can be saved and used in subsequent testing or training.

The Window menu

The **Window** menu lists the current windows that are being displayed by **tlearn** and if a window in the **Window** menu is chosen it is selected as the current window and brought in front of all the other windows.

Dialogue reference

Most of the dialogues used in **tlearn** have been discussed in the corresponding menu item that relates to their use. There remains only the **Training Options** and **Testing Options** dialogues to be discussed in this section. Before these dialogues are discussed some general notes on the dialogues used in **tlearn** are required.

For any entry box on a dialogue that refers to a filename, the following special action is available. If the user double-clicks on the entry box associated with the filename, then a file selection box appears which allows the selection of the appropriate file, or the entry of the (possibly new) file name. This action also ensures that the file that is selected or to be saved is in the correct folder on the file system.

All dialogues have **OK** or **Action** buttons and **Cancel** buttons. Generally the **OK** or action button is the default button and can be selected by pressing Return or Enter. A dialogue can be cancelled by pressing the escape key. When a dialogue is cancelled any of the changes made to the dialogue are also cancelled and the dialogue's items revert to the state they had when the dialogue was brought up. Some of the dialogues have **Dismiss** buttons which allow the dialogue to be dismissed, and the action associated with the dialogue not to be performed, but instead of the dialogue item values reverting to their previous state, the current state of the dialogue items is kept rather than cancelled.

The Training Options dialogue

There are two versions of the **Training Options** dialogue box: small and large. These are shown in Figure B.21. The small **Training Options** dialogue is displayed by default for new projects. The small dialogue allows the modification of a subset of the training options, while the large dialogue gives the user access to all the options that are associated with training. The user can switch between the two dialogues by clicking on the **more...** and **less...** buttons at the bottom left of the dialogues.

The **Training Options** dialogue gives an interface to all of the parameters for all of the training runs performed in the current project. The parameters for different training runs are accessed via the **Prev/Next** buttons at the bottom of the dialogue. For a new project, these buttons are not highlighted. The number of the run and the total number of saved training parameters is given at the top of the dialogue. If the dialogue is showing a set of parameters other than that associated with the latest training run, then a **Remove run** button appears which allows the user to delete previous training run parameters. If the user wants to distinguish a specific training run by giving it a name, then this can be done by double-clicking the top part of the dialogue and the name of the training run can be entered.

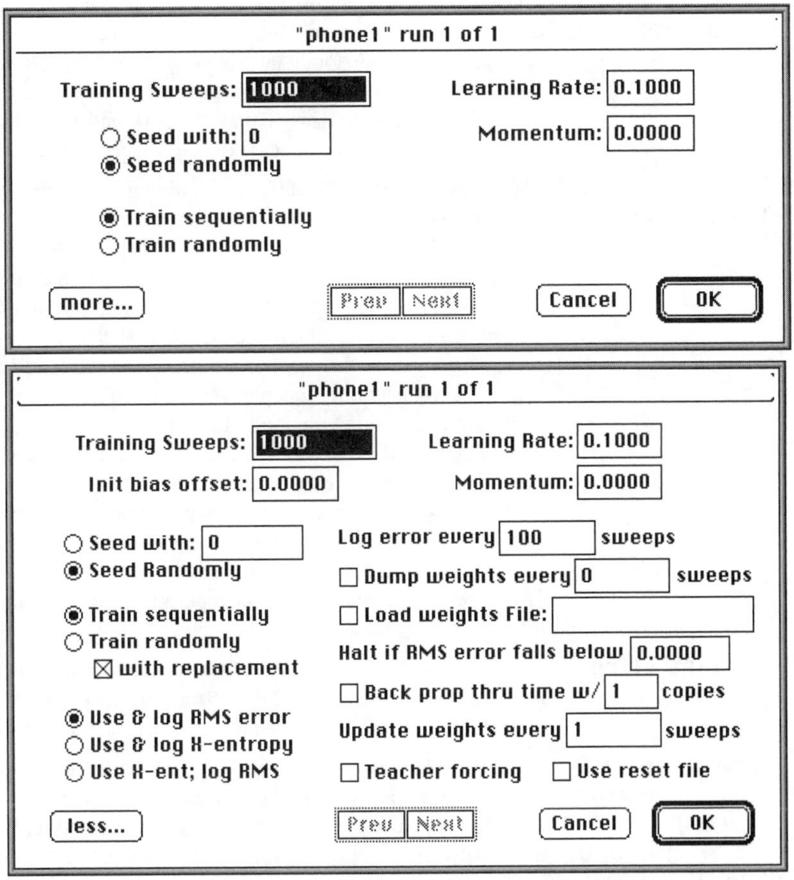

```
                    "phone1" run 1 of 1

    Training Sweeps: [1000]        Learning Rate: [0.1000]

        ○ Seed with: [0]            Momentum: [0.0000]
        ● Seed randomly

        ● Train sequentially
        ○ Train randomly

    ( more... )              [Prev][Next]    ( Cancel )  [[ OK ]]
```

```
                    "phone1" run 1 of 1

    Training Sweeps: [1000]        Learning Rate: [0.1000]
    Init bias offset: [0.0000]        Momentum: [0.0000]

        ○ Seed with: [0]         Log error every [100]    sweeps
        ● Seed Randomly         ☐ Dump weights every [0]    sweeps

        ● Train sequentially    ☐ Load weights File: [          ]
        ○ Train randomly
            ☒ with replacement  Halt if RMS error falls below [0.0000]

        ● Use & log RMS error   ☐ Back prop thru time w/ [1]  copies
        ○ Use & log H-entropy   Update weights every [1]    sweeps
        ○ Use H-ent; log RMS    ☐ Teacher forcing    ☐ Use reset file

    ( less... )              [Prev][Next]    ( Cancel )  [[ OK ]]
```

FIGURE B.21 **Training Options** dialogue boxes

A description of each of the training options is given in the following.

Training Sweeps The number of training sweeps for the training run.

Learning Rate The value of the learning rate for backpropagation. This value is limited between 0.0 and 10.0.

Momentum The value of momentum for backpropgation. Momentum is limited between 0.0 and 1.0.

Seed with:/
Seed randomly

These radio buttons allow the user to specify the initial random seed value (**Seed with:**) or to allow the random seed itself to be chosen randomly. For training runs which are seeded randomly the random seed which was used is displayed in the text box next to **Seed with:**. The random number generator is used for generating the initial random weights for the network and for determining the training data presentation order if the data are to be presented randomly.

Train sequentially/
Train randomly

These radio buttons specify the training data presentation order. If **Train sequentially** is checked, then the training data are presented in the order that appears in the **.data/.teach** files. If **Train randomly** is checked then the training data are presented in a random order. In the large **Training Options** dialogue, the **Train randomly** radio button has an extra check box associated with it, namely **with replacement**. If the **with replacement** box is checked then the training data are presented randomly with replacement, which means that for each subsequent presentation of a training pattern, the next pattern to be presented is chosen randomly from all of the training patterns. When the **with replacement** box is not checked then the first pattern to be presented is chosen randomly from the training patterns but is not replaced; so the next pattern is chosen randomly from the remaining training patterns and is also not replaced. This random choosing continues until there are no more patterns to choose from, at which point all the training patterns are put "back in the hat" to be chosen from again. **Train**ing **randomly** without replacement ensures that all of the training patterns are seen at least once each epoch.

Init bias offset

Allows the setting of an initial offset for the connections from the bias (node 0) to any nodes in the network. This offset only takes effect when the initial random weights of the network are calculated, where it is added to all the bias connections.

Use & log RMS
error
Use & log X-
entropy
Use X-ent; log RMS

These radio buttons give settings for the error function that is used by backpropagation and the error function that is displayed in the error display. Using and logging Root Mean Squared (RMS) error is the default setting. **RMS error** and **Cross entropy** (X-entropy) are often similar in their effects. Cross-entropy is explained in Chapter 9, page 166.

Log error every ..
sweeps

Specifies the sweep interval at which error values are calculated and displayed on the error display.

Dump weights
every .. sweeps

Specifies the sweep interval at which weight files can be saved.

Load weights File

Allows a saved weight file to be used instead of generating initial random weights.

Halt if RMS error
falls below ..

Specifies the error criterion for which training stops.

Back prop thru time w/ .. copies This option is used for training of recurrent networks using the Backpropagation Through Time (BPTT) procedure. Each unit in the network is represented by n multiple copies. Each of the n copies records that unit's activation at time n. Thus, propagating error back through these multiple copies is equivalent to propagating the error back in time. This allows the network to relate error information which comes in at a later time to network states earlier in time.

Update weights every .. sweeps Specifies the sweep interval between weight updates. Online learning occurs when weight updates occur at every sweep. Batch learning occurs when the training presentation is sequential and the update interval is equal to the number of training patterns.

Teacher forcing If set, when output units are fed back (as in Jorda n, 1986 networks), the teacher pattern (target output) will be fed back instead of the actual output. This speeds learning.

Use reset file Causes the use of a `.reset` file during training. A `.reset` file is used for simple recurrent networks to reset the context unit activations.

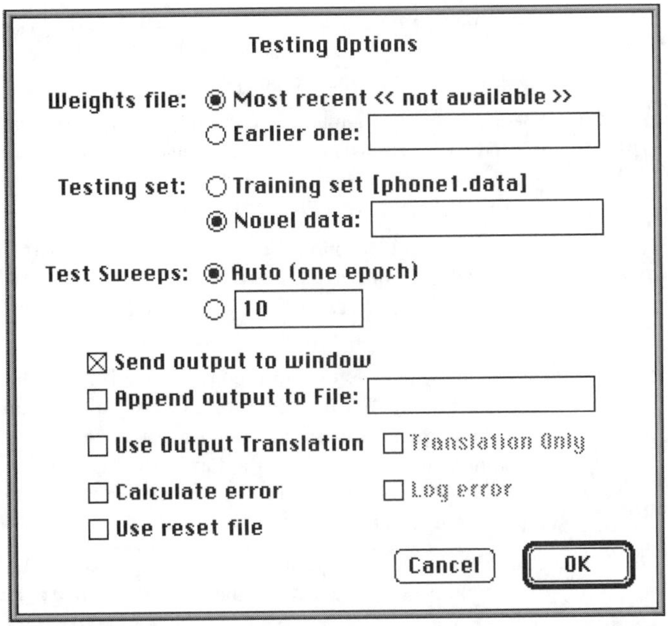

FIGURE B.22 The **Testing Options** dialogue box

The Testing Options dialogue

The **Testing Options** dialogue box is shown in Figure B.22 It is used to change options related to the testing actions (**Verify the network has learned**, **Probe selected nodes**). A description of the options on the dialogue is given in the following.

Weights file   These radio buttons allow the selection of the **Most recent** set of network weights or an **Earlier one:** to be used for testing. The name of the desired weight file can be typed into the text entry box next to the **Earlier one:** choice, or, if the user double-clicks on the text box, a file selection box can be used to select the appropriate file.

Testing set   These radio buttons select the data (and teach) files to be used for testing. Either the **Training set** or a **Novel data:** set can be chosen. As for the earlier weight file selection above, the name of the data set can be typed into the text entry box, or the user can double-click on the text box, and a file selection box can be used to select the `.data` file.

Test Sweeps   The testing action can be set to run for one epoch (that is, a number of sweeps equal to the number of training patterns in the data file) or a specified number of test sweeps entered into the text entry box.

Send output to window/Append output to File   Specify where the output of the testing actions (and also the text output of clustering and **PCA** actions) is sent. The output can be sent to either an **Output** window or appended to a file, specified in the text entry box next to the **Append output to File:** check box, or both.

Use Output Translation   Causes the **Verify** action to use and print the **Output Translation** defined in the **Output Translation** dialogue. The **Translation Only** check box causes the **Verify** action to only output the translation rather than the translation and the output unit activations.

Calculate error   Causes the testing actions to produce network error calculations which appear in the error display. For the calculation of error a `.teach` file is required. For the project's normal training set defined by the `.data` file, there is already a corresponding `.teach` file, but for a novel data set, if an error calculation is required, then a novel `.teach` file must also be present for the testing actions to produce a valid error calculation.

Log error   Causes the error calculation if specified by the **Calculate error** check box to be written to a file. The name of the file is `<project_name>.err`

Use reset file   This check box causes the use of a `.reset` file during testing actions. A `.reset` file is mostly used for simple recurrent networks which require the activations of context units to be reset.

Command key and shortcut reference

Menu command keys

Menu keys can be seen in the appropriate menus where, if the menu item has an associated command key, it is shown at the right of the menu item.

Windows	Mac	Action
Ctrl+A	⌘-A	Edit/Select All
Ctrl+C	⌘-C	Edit/Copy
Ctrl+E	⌘-E	Search/Enter Selection
Ctrl+F	⌘-F	Search/Find…
F3	⌘-G	Search/Find Again
F4	⌘-H	Search/Replace & Find Again
Ctrl+J	⌘-J	Network/Testing Options…
Ctrl+K	⌘-K	Network/Verify the network has learned
Ctrl+L	⌘-L	Network/Probe selected nodes
Ctrl+N	⌘-N	File/New
Ctrl+O	⌘-O	File/Open…
Ctrl+P	⌘-P	File/Print…
Alt-F-x	⌘-Q	File/Quit
Ctrl+R	⌘-R	Network/Resume training
Ctrl+S	⌘-S	File/Save
Ctrl+T	⌘-T	Network/Train the network
Ctrl+U	⌘-U	Network/Abort
Ctrl+V	⌘-V	Edit/Paste
Alt-F-C	⌘-W	File/Close
Ctrl+X	⌘-X	Edit/Cut
Ctrl+Y	⌘-Y	Network/Training Options…
Ctrl+Z	⌘-Z	Edit/Undo
Ctrl+H	⌘-=	Search/Replace
Ctrl+G	⌘-`	Search/Go To Line…

Editor quick keys

Shift-arrow-keys allows the selection of text (or, if text is already selected, the extension or reduction of the text selection) to be done from the keyboard rather than with a click-drag action of the mouse.

`Ctrl-left, Ctrl-right`	moves the cursor to the next or previous word boundary
`Ctrl-up, Ctrl-down`	moves the cursor up/down a screenful of text
`⌘-left, ⌘-right`	moves the cursor to the beginning/end of a line
`⌘-up, ⌘-down`	moves the cursor to the beginning/end of a document

Troubleshooting

This section lists common problems and solutions.

Problem: Can't lesion weights file. **Lesion** command is disabled.

Solution: Before the lesioning action can be performed, a project must be specified. The reason for this is that the network configuration must be known for lesioning to be performed.

Problem: I have specified an **Output Translation** file in the **Special** menu, but when I probe or verify the output translation doesn't appear.

Solution: In the **Testing Options** dialogue the **Use Output Translation** check box must be set for the **Output Translation** to be used. If you only want the **Output Translation** and not the output node activations, then the **Output Translation Only** box should be checked.

Problem: I want to display principal components other than the first two or three. How do I do this?

Solution: Specify a subset in the appropriate part of the **Principal Components Analysis** dialogue box.

Problem: Why doesn't the project change even though I close the old `.cf`, `.data`, `.teach` files and open the new files?

Solution: To use a different set of project files you need to create or open the project using the **New Project...**/**Open Project...** commands in the **Network** menu. The **Status** Display shows the name of the current project.

Problem: The **Translate...** action translates too much and translates things I don't want translated.

Solution: The order of things in the translate file is important. Specifically, care must be taken that a later translation doesn't inadvertently retranslate a previous translation. This could occur if letters were translated to numerical vectors, and then a later translation action translated digits to letters. Unless this was a desired effect, this double translation will possibly cause havoc to the translation as intended. One piece of advice here is to ensure that no strings on the left of a translation rule appear on the right of a translation rule.

Problem: When I run **tlearn** on several projects, I'm told that there is not enough memory (for **data** or **teach** files, for example).

Solution: After running consecutive projects, **tlearn** may fail to release all the memory associated with the files it has used. Quitting **tlearn** and restarting should solve the problem.

References

Berko, J. (1958). The child's learning of English morphology. *Word, 14*, 150-177.

Elman, J.L. (1990). Finding structure in time. *Cognitive Science, 14*, 179-211.

Elman, J.L. (1993). Learning and development in neural networks: The importance of starting small. *Cognition, 48*, 71-99.

Elman, J. L., Bates, E. A., Johnson, M. H., Karmiloff-Smith, A., Parisi, D. & Plunkett, K. (1996). *Rethinking Innateness: A connectionist perspective on development*. Cambridge, MA: MIT Press.

Ervin, S. (1964). Imitation and structural change in children's language. In E.H. Lenneberg (Ed.), *New Directions in the Study of Language*. Cambridge, MA: MIT Press.

Fodor, J.A., & Pylyshyn, Z.W. (1988). Connectionism and cognitive architecture: A critical analysis. In S. Pinker & J. Mehler (Eds.), *Connections and Symbol*s (pp. 3-71). *(Cognition* Special Issue). Cambridge, MA: MIT Press/Bradford Books.

Gold, E.M. (1967). Language identification in the limit. *Information and Control, 16*, 447-474.

Hertz, J.A., Krøgh, A., & Palmer, R.G. (1991). *Introduction to the theory of neural computation*. Lecture Notes Volume I, Santa Fe Institute, Studies in the Sciences of Complexity. Redwood City, CA: Addison-Wesley.

Hinton, G. E., (1986) Learning Distributed Representations of Concepts. In *Proceedings of the Eighth Annual Conference of the Cognitive Science Society* (Amherst), 1–12, Hillsdale: Erlbaum

Inhelder, B., & Piaget, J. (1958). *The Growth of Logical Thinking from Childhood to Adolescence*. New York: Basic Books.

Jordan, M. (1986). Serial order: A parallel distributed processing approach. Technical Report 8604 . San Diego: Institute for Cognitive Science. University of California.

Le Cun, Y. (1985) Une procédure d'apprentissage pour réseau à seuil assymétrique [A learning procedure for an assymetric threshold network] *Proceedings of Cognitiva, i5*, 599–604. Paris.

Lehky, S.R., & Sejnowski, T.J. (1988). Network model of shape-from-shading: Neural function arises from both receptive and projective fields. *Nature, 333*, 452-454.

Marcus, G., Ullman, M., Pinker, S., Hollander, M., Rosen, T.J., & Xu, F. (1992). Overregularization in language acquisition. *Monographs of the Society for Research in Child Development, 57.*

McClelland, J.L., (1989). Parallel distributed processing: Implications for cognition and development. *In R.G.M. Morris (Ed.), Parallel Distributed Processing: Implications for Psychology and Neurobiology* (pp. 9-45). Oxford: Clarendon Press.

Mozer, M. & Smolensky, P. (1989) Using Relevance to Reduce Network Size Automatically, *Connection Science, 1*, 3–17

Newport, E.L. (1988). Constraints on learning and their role in language acquisition: Studies of the acquisition of American Sign Language. *Language Sciences, 10,* 147-172.

Newport, E.L. (1990). Maturational constraints on language learning. *Cognitive Science, 14,* 11-28.

Pinker, S., & Prince, A. (1988). On language and connectionism: Analysis of a parallel distributed processing model of language acquisition. *Cognition, 28*, 73-193.

Plunkett, K., & Marchman, V. (1991). U-shaped learning and frequency effects in a multi-layered perceptron: Implications for child language acquisition. *Cognition, 38*, 43-102.

Plunkett, K., & Marchman, V. (1993). From rote learning to system building: Acquiring verb morphology in children and connectionist nets. *Cognition , 48,* 21-69.

Plunkett, K., & Marchman, V. (1996). Learning from a connectionist model of the acquisition of the English past tense. *Cognition , 61,* 299–308.

Rumelhart, D.E., Hinton, G., & Williams, R. (1986). Learning internal representations by error propagation. In D.E. Rumelhart & J.L. McClelland (Eds.), *Parallel Distributed Processing: Explorations in the Microstructure of Cognition: Vol. 1. Foundations* (pp. 318-

362). Cambridge, MA: MIT Press.

Rumelhart, D.E., & McClelland, J.L. (1986). On learning the past tenses of English verbs. In D.E. Rumelhart & J.L. McClelland (Eds.), *Parallel distributed processing: Explorations in the microstructure of cognition. Volume 2. Psychological and biological models* (pp. 216-271). Cambridge, MA: MIT Press.

Sanger, T. D. (1989) Contribution analysis: A technique for assigning responsibilities to hidden units in connectionsit networks. *Connection Science, 1,* 115–138.

Siegler, R. (1981). Developmental sequences within and between concepts. *Monographs of the Society for Research in Child Development, 46,* Whole No. 2.

Werbos, P. (1974) Beyond Regression: New Tools for Prediction and Analysis in the Behavioural Sciences. Ph.D. Thesis, Harvard University.

Index

All commands, menu selections, dialogue box options and display options are listed in **bold** type.